In the Heart of Africa

by Samuel White Baker

CONTENTS.

CHAPTER I.

The Nubian desert--The bitter well--Change of plans--An irascible dragoman--Pools of the Atbara--One secret of the Nile--At Cassala

CHAPTER II.

Egypt's rule of the Soudan--Corn-grinding in the Soudan--Mahomet meets relatives--The parent of Egypt--El Baggar rides the camel

CHAPTER III.

The Arabs' exodus--Reception by Abou Sinn--Arabs dressing the hair--Toilet of an Arab woman--The plague of lice--Wives among the Arabs--The Old Testament confirmed

CHAPTER IV.

On the Abyssinian border--A new school of medicine--Sacred shrines and epidemics

CHAPTER V.

A primitive craft--Stalking the giraffes--My first giraffes-Rare sport with the finny tribe--Thieving elephants

CHAPTER VI.

Preparations for advance--Mek Nimmur makes a foray--The Hamran elephant-hunters--In the haunts of the elephant-- A desperate charge

CHAPTER VII.

The start from Geera--Feats of horsemanship--A curious chase-- Abou Do wins a race--Capturing a young buffalo--Our island camp--Tales of the Base

CHAPTER VIII.

The elephant trumpets--Fighting an elephant with swords-- The forehead-shot--Elephants in a panic--A superb old Neptune--The harpoon reaches its aim--Death of the hippopotamus--Tramped by an elephant

CHAPTER IX.

Fright of the Tokrooris--Deserters who didn't desert--Arrival of the Sherrif brothers--Now for a tally-ho!--On the heels of the rhinoceroses--The Abyssinian rhinoceros--Every man for himself

CHAPTER X.

A day with the howartis--A hippo's gallant fight--Abou Do leaves us--Three yards from a lion--Days of delight--A lion's furious rage--Astounding courage of a horse

CHAPTER XI.

The bull-elephant--Daring Hamrans--The elephant helpless--Visited by a minstrel--A determined musician--The nest of the outlaws-- The Atbara River

CHAPTER XII.

Abyssinian slave-girls--Khartoum--The Soudan under Egyptian rule-- Slave-trade in the Soudan--The obstacles ahead

CHAPTER XIII.

Gondokoro--A mutiny quelled--Arrival of Speke and Grant--The sources of the Nile-Arab duplicity--The boy-slave's story--Saat adopted

CHAPTER XIV.

Startling disclosures--The last hope seems gone--The Bari chief's advice--Hoping for the best--Ho for Central Africa!

CHAPTER XV.

A start made at last--A forced march--Lightening the ship--Waiting for the caravan--Success hangs in the balance--The greatest rascal in Central Africa--Legge demands another bottle

CHAPTER XVI.

The greeting of the slave-traders--Collapse of the mutiny--African funerals-Visit from the Latooka chief--Bokke makes a suggestion-- Slaughter of the Turks--Success as a prophet--Commoro's philosophy

CHAPTER XVII.

Disease in the camp--Forward under difficulties--Our cup of misery overflows--A rain-maker in a dilemma-Fever again--Ibrahim's quandary-Firing the prairie

CHAPTER XVIII.

Greeting from Kamrasi's people--Suffering from the sins of others-Alone among savages--The free-masonry of Unyoro.--Pottery and civilization

CHAPTER XIX.

Kamrasi's cowardice--Interview with the king--The exchange of blood-- The rod beggar's last chance--An astounded sovereign

CHAPTER XX.

A satanic escort--Prostrated by sun-stroke--Days and nights of sorrow--The reward for all our labor

CHAPTER XXI.

The cradle of the Nile--Arrival at Magungo--The blind leading the blind--Murchison Falls

CHAPTER XXII.

Prisoners on the island--Left to starve--Months of helpless- ness--We rejoin the Turks--The real Kamrasi--In the presence of royalty

CHAPTER XXIII.

The hour of deliverance--Triumphal entry into Gondokoro-- Homeward bound--The plague breaks out--Our welcome at Khartoum--Return to civilization

IN THE HEART OF AFRICA.

CHAPTER I.

The Nubian desert--The bitter well--Change of plans--An irascible dragoman--Pools of the Atbara--One secret of the Nile--At Cassala.

In March, 1861, I commenced an expedition to discover the sources of the Nile, with the hope of meeting the East African expedition of Captains Speke and Grant, that had been sent by the English Government from the South via Zanzibar, for the same object. I had not the presumption to publish my intention, as the sources of the Nile had hitherto defied all explorers, but I had inwardly determined to accomplish this difficult task or to die in the attempt. From my youth I had been inured to hardships and endurance in wild sports in tropical climates, and when I gazed upon the map of Africa I had a wild hope, mingled with humility, that, even as the insignificant worm bores through the hardest oak, I might by perseverance reach the heart of Africa.

I could not conceive that anything in this world has power to resist a determined will, so long as health and life remain. The failure of every former attempt to reach the Nile source did not astonish me, as the expeditions had consisted of parties, which, when difficulties occur, generally end in difference of opinion and in retreat; I therefore determined to proceed alone, trusting in the guidance of a Divine Providence and the good fortune that sometimes attends a tenacity of purpose. I weighed carefully the chances of the undertaking. Before me, untrodden Africa; against me, the obstacles that had defeated the world since its creation; on my side, a somewhat tough constitution, perfect independence, a long experience in savage life, and both time and means, which I intended to devote to the object without limit.

England had never sent an expedition to the Nile sources previous to that under the command of Speke and Grant. Bruce, ninety years before, had succeeded in tracing the source of the Blue or Lesser Nile; thus the honor of that discovery belonged to Great Britain. Speke was on his road from the South, and I felt confident that my gallant friend would leave his bones upon the path rather than submit to failure. I trusted that England would not be beaten, and although I hardly dared to hope that I could succeed where others greater than I had failed, I determined to sacrifice all in the attempt.

Had I been alone, it would have been no hard lot to die upon the untrodden path before me; but there was one who, although my greatest comfort, was also my greatest care, one whose life yet dawned at so early an age that womanhood was still a future. I shuddered at the prospect for her, should she be left alone in savage lands at my death; and gladly would I have left her in the luxuries of home instead of exposing her to the miseries of Africa. It was in vain that I implored her to remain, and that I painted the difficulties and perils still blacker than I supposed they really would be. She was resolved, with woman's constancy and devotion, to share all dangers and to follow me through each rough footstep of the wild life before me. "And Ruth said, Entreat me not to leave thee, or to return from following after thee; for whither thou goest I will go, and where thou lodgest I will lodge; thy people shall be my people, and thy God my God; where thou diest will I die, and there will I be buried: the Lord do so to me, and more also, if aught but death part thee and me."

Thus accompanied by my wife, on the 15th of April, 1861, I sailed up the Nile from Cairo. The wind blew fair and strong from the north, and we flew toward the south against the stream, watching those mysterious waters with a firm resolve to track them to their distant fountain.

I had a firman from the Viceroy, a cook, and a dragoman. Thus my impedimenta were not numerous. The firman was an order to all Egyptian officials for assistance; the cook was dirty and incapable; and the interpreter was nearly ignorant of English, although a professed polyglot. With this small beginning, Africa was before me, and thus I commenced the search for the sources of the Nile.

On arrival at Korosko, twenty-six days from Cairo, we started across the Nubian Desert. During the cool months, from November until February, the desert journey is not disagreeable; but the vast area of glowing sand exposed to the scorching sun of summer, in addition to the withering breath of the simoom, renders the forced march of two hundred and thirty miles in seven days, at two and a half miles per hour, one of the most fatiguing journeys that can he endured.

We entered a dead level plain of orange-colored sand, surrounded by pyramidical hills. The surface was strewn with objects resembling cannon shot and grape of all sizes from a 32-pounder downward, and looked like the old battle-field of some infernal region--rocks glowing with heat, not a vestige of vegetation, barren, withering desolation. The slow rocking step of the camels was most irksome, and, despite the heat, I dismounted to examine the Satanic bombs and cannon shot. Many of them were as perfectly round as though cast in a mould, others were egg-shaped, and all were hollow. With some difficulty I broke them, and found them to contain a bright red sand. They were, in fact, volcanic bombs that had been formed by the ejection of molten lava to a great height from active volcanoes; these had become globular in falling, and, having cooled before they reached the earth, they retained their forms as hard spherical bodies, precisely resembling cannon shot. The exterior was brown, and appeared to be rich in iron. The smaller specimens were the more perfect spheres, as they cooled quickly; but many of the heavier masses had evidently reached the earth when only half solidified, and had collapsed upon falling. The sandy plain was covered with such vestiges of volcanic action, and the infernal bombs lay as imperishable relics of a hailstorm such as may have destroyed Sodom and Gomorrah.

Passing through this wretched solitude, we entered upon a scene of surpassing desolation. Far as the eye could reach were waves like a stormy sea, gray, coldlooking waves in the burning heat; but no drop of water. It appeared as though a sudden curse had turned a raging sea to stone. The simoom blew over this horrible wilderness, and drifted the hot sand into the crevices of the rocks, and the camels drooped their heads before the suffocating wind; but still the caravan noiselessly crept along over the rocky undulations, until the stormy sea was passed; once more we were upon a boundless plain of sand and pebbles.

In forty-six hours and forty-five minutes' actual marching from Korosko, we reached Moorahd, "the bitter well." This is a mournful spot, well known to the tired and thirsty camel, the hope of reaching which has urged him fainting on his weary way to drink one draught before he dies. This is the camel's grave. Situated half way between Korosko and Abou Hammed, the well of Moorahd is in an extinct crater, surrounded upon all sides but one by precipitous cliffs about three hundred feet high. The bottom is a dead flat, and forms a valley of sand about two hundred and fifty yards wide. In this bosom of a crater, salt and bitter water is found at a depth of only six feet from the surface. To this our tired camels frantically rushed upon being unloaded.

The valley was a "valley of dry bones." Innumerable skeletons of camels lay in all directions-the ships of the desert thus stranded on their voyage. Withered heaps of parched skin and bone lay here and there, in the distinct forms in which the camels had gasped their last. The dry desert air had converted the hide into a coffin. There were no flies here, thus there were no worms to devour the carcasses ; but the usual sextons were the crows, although sometimes too few to perform their office. These were perched upon the overhanging cliffs ; but no sooner had our overworked camels taken their long draught and lain down exhausted on the sand, than by common consent they descended from their high places and walked round and round each tired beast.

As many wretched animals simply crawl to this spot to die, the crows, from long experience and constant practice, can form a pretty correct diagnosis upon the case of a sick camel. They had evidently paid a professional visit to my caravan, and were especially attentive in studying the case of one particular camel that was in a very weakly condition and had stretched itself full length upon the sand; nor would they leave it until it was driven forward.

Many years ago, when the Egyptian troops first conquered Nubia, a regiment was destroyed by thirst in crossing this desert. The men, being upon a limited allowance of water, suffered from extreme thirst, and deceived by the appearance of a mirage that exactly resembled a beautiful lake, they insisted on being taken to its banks by the Arab guide. It was in vain that the guide assured them that the lake was unreal, and he refused to lose the precious time by wandering from his course. Words led to blows, and he was killed by the soldiers, whose lives depended upon his guidance. The whole regiment turned from the track and rushed toward the welcome waters. Thirsty and faint, over the burning sands they hurried; heavier and heavier their footsteps became; hotter and hotter their breath, as deeper they pushed into the desert, farther and farther from the lost track where the pilot lay in his blood; and still the mocking spirits of the desert, the afreets of the mirage, led them on, and the hike glistening in the sunshine tempted them to bathe in its cool waters, close to their eyes, but never at their lips. At length the delusion vanished--the fatal lake had turned to burning sand! Raging thirst and horrible despair! the pathless desert and the murdered guide! lost! lost! all lost! Not a man ever left the desert, but they were subsequently discovered, parched and withered corpses, by the Arabs sent upon the search.

During our march the simoom was fearful, and the heat so intense that it was impossible to draw the guncases out of their leather covers, which it was necessary to cut open. All woodwork was warped; ivory knife-handles were split; paper broke when crunched in the hand, and the very marrow seemed to he dried out of the bones. The extreme dryness of the air induced an extraordinary amount of electricity in the hair and in all woollen materials. A Scotch plaid laid upon a blanket for a few hours adhered to it, and upon being withdrawn at night a sheet of flame was produced, accompanied by tolerably loud reports.

We reached Berber on May 31st, and spent a week in resting after our formidable desert march of fifteen days. From the slight experience I had gained in the journey, I felt convinced that success in my Nile expedition would be impossible without a knowledge of Arabic. My dragoman had me completely in his power, and I resolved to become independent of all interpreters as soon as possible. I therefore arranged a plan of exploration for the first year, to embrace the affluents to the Nile from the Abyssinian range of mountains, intending to follow up the Atbara River from its junction with the Nile in latitude 17 deg. 37 min. (twenty miles south of Berber), and to examine all the Nile tributaries from the southeast as far as the Blue Nile, which river I hoped ultimately to descend to Khartoum. I imagined that twelve months would be sufficient to complete such an exploration, by which time I should have gained a sufficient knowledge of the Arabic to render me able to converse fairly well.

The wind at this season (June) was changeable, and strong blasts from the south were the harbingers of the approaching rainy season. We had no time to lose, and we accordingly arranged to start. I discharged my dirty cook, and engaged a man who was brought by a coffeehouse keeper, by whom he was highly recommended; but, as a precaution against deception, I led him before the Mudir, or Governor, to be registered before our departure. To my astonishment, and to his infinite disgust, he was immediately recognized as an old offender, who had formerly been imprisoned for theft! The Governor, to prove his friendship and his interest in my welfare, immediately sent the police to capture the coffee-house keeper who had recommended the cook. No sooner was the unlucky surety brought to the Divan than he was condemned to receive two hundred lashes for having given a false character. The sentence was literally carried out, in spite of my remonstrance, and the police were ordered to make the case public to prevent a recurrence. The Governor assured me that, as I held a firman from the Viceroy, he could not do otherwise, and that I must believe him to be my truest friend. "Save me from my friends," was an adage quickly proved. I could not procure a cook nor any other attendant, as every one was afraid to guarantee a character, lest he might come in for his share of the two hundred lashes!

The Governor came to my rescue, and sent immediately the promised Turkish soldiers, who were to act in the double capacity of escort and servants. They were men of totally opposite characters. Hadji Achmet was a hardy, powerful, dare-devil-looking Turk, while Hadji Velli was the perfection of politeness, and as gentle as a lamb. My new allies procured me three donkeys in addition to the necessary baggage camels, and we started from Berber on the evening of the 10th of June for the junction of the Atbara River with the Nile.

Mahomet, Achmet, and Ali are equivalent to Smith, Brown, and Thompson. Accordingly, of my few attendants, my dragoman was Mahomet, and my principal guide was Achmet, and subsequently I had a number of Alis. Mahomet was a regular Cairo dragoman, a native of Dongola, almost black, but exceedingly tenacious regarding his shade of color, which he declared to be light brown. He spoke very bad English, was excessively conceited, and irascible to a degree. He was one of those dragomans who are accustomed to the civilized expeditions of the British tourist to the first or second cataract, in a Nile boat replete with conveniences and luxuries, upon which the dragoman is monarch supreme, a whale among the minnows, who rules the vessel, purchases daily a host of unnecessary supplies, upon which he clears his profit, until he returns to Cairo with his pockets filled sufficiently to support him until the following Nile season. The short three months' harvest, from November until February, fills his granary for the year. Under such circumstances the temper should be angelic.

But times had changed. To Mahomet the very idea of exploration was an absurdity. He had never believed in it front the first, and he now became impressed with the fact that he was positively committed to an undertaking that would end most likely in his death, if not in terrible difficulties; he determined, under the circumstances, to make himself as disagreeable as possible to all parties. With this amiable resolution he adopted a physical infirmity in the shape of deafness. In reality, no one was more acute in hearing, but as there are no bells where there are no houses, he of course could not answer such a summons, and he was compelled to attend to the call of his own name--"Mahomet! Mahomet!" No reply, although the individual were sitting within a few feet, apparently absorbed in the contemplation of his own boots. "MaHOMet!" with an additional emphasis upon the second syllable. Again no response. "Mahomet, you rascal, why don't you answer?" This energetic address would effect a change in his position. The mild and lamb-like dragoman of Cairo would suddenly start from the ground, tear his own hair from his head in handfuls, and shout, "Mahomet! Mahomet! Mahomet! always Mahomet! D--n Mahomet! I wish he were dead, or back in Cairo, this brute Mahomet!" The irascible dragoman would then beat his own head unmercifully with his fists, in a paroxysm of rage.

To comfort him I could only exclaim, "Well done, Mahomet! thrash him; pommel him well; punch his head; you know him best; he deserves it; don't spare him!" This advice, acting upon the natural perversity of his disposition, generally soothed him, and he ceased punching his head. This man was entirely out of his place, if not out of his mind, at certain moments, and having upon one occasion smashed a basin by throwing it in the face of the cook, and upon another occasion narrowly escaped homicide by throwing an axe at a man's head, which missed by an inch, he became a notorious character in the little expedition.

We left Berber in the evening, and about two hours after sunset of the following day reached the junction of the Nile and Atbara. The latter presented a curious appearance. In no place was it less than four hundred yards in width, and in many places much wider. The banks were from twenty-five to thirty feet deep, and had evidently been overflowed during floods; but now the river bed was dry sand, so glaring that the sun's reflection was almost intolerable. The only shade was afforded by the evergreen dome palms; nevertheless the Arabs occupied the banks at intervals of three or four miles, wherever a pool of water in some deep bend of the dried river's bed offered an attraction. In such places were Arab villages or camps, of the usual mat tents formed of the dome- palm leaves.

Many pools were of considerable size and of great depth. In flood-time a tremendous torrent sweeps down the course of the Atbara, and the sudden bends of the river are hollowed out by the force of the stream to a depth of twenty or thirty feet below the level of the bed. Accordingly these holes become reservoirs of water when the river is otherwise exhausted. In such asylums all the usual inhabitants of this large river are crowded together in a comparatively narrow space. Although these pools vary in size, from only a few hundred yards to a mile in length, they are positively full of life; huge fish, crocodiles of immense size, turtles, and occasionally hippopotami, consort together in close and unwished-for proximity. The animals of the desert-- gazelles, hyenas, and wild asses--are compelled to resort to these crowded drinking-places, occupied by the flocks of the Arabs equally with the timid beasts of the chase. The birds that during the cooler months would wander free throughout the country are now collected in vast numbers along the margin of the exhausted river; innumerable doves, varying in species, throng the trees and seek the shade of the dome-palms; thousands of desert grouse arrive morning and evening to drink and to depart; while birds in multitudes, of lovely plumage, escape from the burning desert and colonize the poor but welcome bushes that fringe the Atbara River.

After several days' journey along the bank of the Atbara we halted at a spot called Collodabad, about one hundred and sixty miles from the Nile junction. A sharp bend of the river had left a deep pool about a mile in length, and here a number of Arabs were congregated, with their flocks and herds.

On the evening of June 23d I was lying half asleep upon my bed by the margin of the river, when I fancied that I heard a rumbling like distant thunder. I had not heard such a sound for months, but a low, uninterrupted roll appeared to increase in volume, although far distant. Hardly had I raised my head to listen more attentively when a confusion of voices arose from the Arabs' camp, with a sound of many feet, and in a few minutes they rushed into my camp, shouting to my men in the darkness, "El Bahr! El Bahr!" (the river! the river!)

We were up in an instant, and my interpreter, Mahomet, in a state of intense confusion, explained that the river was coming down, and that the supposed distant thunder was the roar of approaching water.

Many of the people were asleep on the clean sand on the river's bed; these were quickly awakened by the Arabs, who rushed down the steep bank to save the skulls of two hippopotami that were exposed to dry. Hardly had they descended when the sound of the river in the darkness beneath told us that the water had arrived, and the men, dripping with wet, had just sufficient time to drag their heavy burdens up the bank.

All was darkness and confusion, everybody talking and no one listening; but the great event had occurred; the river had arrived "like a thief in the night". On the morning of the 24th of June, I stood on the banks of the noble Atbara River at the break of day. The wonder of the desert! Yesterday there was a barren sheet of glaring sand, with a fringe of withered bushes and trees upon its borders, that cut the yellow expanse of desert. For days we had journeyed along the exhausted bed; all Nature, even in Nature's poverty, was most poor: no bush could boast a leaf, no tree could throw a shade, crisp gums crackled upon the stems of the mimosas, the sap dried upon the burst bark, sprung with the withering heat of the simoom. In one night there was a mysterious change. Wonders of the mighty Nile! An army of water was hastening to the wasted river. There was no drop of rain, no thunder-cloud on the horizon to give hope. All had been dry and sultry, dust and desolation yesterday; to-day a magnificent stream, some five hundred yards in width and from fifteen to twenty feet in depth, flowed through the dreary desert! Bamboos and reeds, with trash of all kinds, were hurried along the muddy waters. Where were all the crowded inhabitants of the pool? The prison doors were broken, the prisoners were released, and rejoiced in the mighty stream of the Atbara.

The 24th of June, 1861, was a memorable day. Although this was actually the beginning of my work, I felt that by the experience of this night I had obtained a clew to one portion of the Nile mystery, and that, as "coming events cast their shadows before," this sudden creation of a river was but the shadow of the great cause. The rains were pouring in Abyssinia! THESE WERE SOURCES OF THE NILE!

The journey along the margin of the Atbara was similar to the route from

Berber, through a vast desert, with a narrow band of trees that marked the course of the river. The only change was the magical growth of the leaves, which burst hourly from the swollen buds of the mimosas. This could be accounted for by the sudden arrival of the river, as the water percolated rapidly through the sand and nourished the famishing roots.

At Gozerajup, two hundred and forty-six miles from Berber, our route was changed. We had hitherto followed the course of the Atbara, but we were now to leave that river on our right, while we travelled about ninety miles south-east to Cassala, the capital of the Taka country, on the confines of Abyssinia, and the great depot for Egyptian troops.

The entire country from Gozerajup to Cassala is a dead flat, upon which there is not one tree sufficiently large to shade a full-sized tent. There is no real timber in the country; but the vast level extent of soil is a series of open plains and low bush of thorny mimosa. There is no drainage upon this perfect level; thus, during the rainy season, the soakage actually melts the soil, and forms deep holes throughout the country, which then becomes an impenetrable slough, bearing grass and jungle. No sooner had we arrived in the flooded country than my wife was seized with a sudden and severe fever, which necessitated a halt upon the march, as she could no longer sit upon her camel. In the evening several hundreds of Arabs arrived and encamped around our fire. It was shortly after sunset, and it was interesting to watch the extreme rapidity with which these swarthy sons of the desert pitched their camp. A hundred fires were quickly blazing; the women prepared the food, and children sat in clusters around the blaze, as all were wet from paddling through the puddled ground from which they were retreating.

No sooner was the bustle of arrangement completed than a gray old man stepped forward, and, responding to his call, every man of the hundreds present formed in line, three or four deep. At once there was total silence, disturbed only by the crackling of the fires or by the cry of a child; and with faces turned to the east, in attitudes of profound devotion, the wild but fervent followers of Mahomet repeated their evening prayer. The flickering red light of the fires illumined the bronze faces of the congregation, and as I stood before the front line of devotees, I tools off my cap in respect for their faith, and at the close of their prayer made my salaam to their venerable Faky (priest); he returned the salutation with the cold dignity of an Arab.

On the next day my wife's fever was renewed, but she was placed on a dromedary and we reached Cassala about sunset. The place is rich in hyenas, and the night was passed in the discordant howling of these disgusting but useful animals. They are the scavengers of the country, devouring every species of filth and clearing all carrion from the earth. Without the hyenas and vultures the neighborhood of a Nubian village would be unbearable. It is the idle custom of the people to leave unburied all animals that die; thus, among the numerous flocks and herds, the casualties would create a pestilence were it not for the birds and beasts of prey.

On the following morning the fever had yielded to quinine, and we were enabled to receive a round of visits --the governor and suite, Elias Bey, the doctor and a friend, and, lastly, Malem Georgis, an elderly Greek merchant, who, with great hospitality, insisted upon our quitting the sultry tent and sharing his own roof. We therefore became his guests in a most comfortable house for some days. Here we discharged our camels, as our Turk, Hadji Achmet's, service ended at this point, and proceeded to start afresh for the Nile tributaries of Abyssinia.

CHAPTER II.

Egypt's rule of the Soudan--Corn-grinding in the Soudan--Mahomet meets relatives--The parent of Egypt--El Baggar rides the camel.

Cassala was built about twenty years before I visited the country, after Taka had been conquered and annexed to Egypt. The general annexation of the Soudan and tile submission of the numerous Arab tribes to the Viceroy have been the first steps necessary to the improvement of the country. Although the Egyptians are hard masters, and do not trouble themselves about the future well-being of the conquered races, it must be remembered that, prior to the annexation, all the tribes were at war among themselves. There was neither government nor law; thus the whole country was closed to Europeans. At the time of my visit to Cassala in 1861 the Arab tribes were separately governed by their own chiefs or sheiks, who were responsible to the Egyptian authorities for the taxes due from their people. Since that period the entire tribes of all denominations have been placed under the authority of that grand old Arab patriarch, Achmet Abou Sinn, to be hereafter mentioned. The iron hand of despotism has produced a marvellous change among the Arabs, who are rendered utterly powerless by the system of government adopted by the Egyptians; unfortunately, this harsh system has the effect of paralyzing all industry.

The principal object of Turks and Egyptians in annexation is to increase their power of taxation by gaining an additional number of subjects. Thus, although many advantages have accrued to the Arab provinces of Nubia through Egyptian rule, there exists very much mistrust between the governed and the governing. Not only are the camels, cattle, and sheep subjected to a tax, but every attempt at cultivation is thwarted by the authorities, who impose a fine or tax upon the superficial area of the cultivated land. Thus, no one will cultivate more than is absolutely necessary, as he dreads the difficulties that broad acres of waving crops would entail upon his family. The bona fide tax is a bagatelle to the amounts squeezed from him by the extortionate soldiery, who are the agents employed by the sheik; these must have their share of the plunder, in excess of the amount to be delivered to their employer; he also must have his plunder before he parts with the bags of dollars to the governor of the province. Thus the unfortunate cultivator is ground down. Should he refuse to pay the necessary "backsheesh" or present to the tax-collectors, some false charge is trumped up against him, and he is thrown into prison. As a green field is an attraction to a flight of locusts in their desolating voyage, so is a luxuriant farm in the Soudan a point for the tax-collectors of Upper Egypt. I have frequently ridden several days' journey through a succession of empty villages, deserted by the inhabitants upon the report of the soldiers' approach. The women and children, goats and cattle, camels and asses, had all been removed into the wilderness for refuge, while their crops of corn had been left standing for the plunderers, who would be too idle to reap and thrash the grain.

Notwithstanding the miserable that fetters the steps of improvement, Nature has bestowed such great capabilities of production in the fertile soil of this country that the yield of a small surface is more than sufficient for the requirements of the population, and actual poverty is unknown. The average price of dhurra is fifteen piastres per "rachel," or about 3s. 2d. for five hundred pounds upon the spot where it is grown. The dhurra (Sorghum andropogon) is the grain most commonly used throughout the Soudan; there are great varieties of this plant, of which the most common are the white and the red. The land is not only favored by Nature by its fertility, but the intense heat of the summer is the laborer's great assistant. As before described, all vegetation entirely disappears in the glaring sun, or becomes so dry that it is swept off by fire; thus the soil is perfectly clean and fit for immediate cultivation upon the arrival of the rains.

The tool generally used is similar to the Dutch hoe. With this simple implement the surface is scratched to the depth of about two inches, and the seeds of the dhurra are dibbled in about three feet apart, in rows from four to five feet in width. Two seeds are dropped into each hole. A few days after the first shower they rise above the ground, and when about six inches high the whole population turn out of their villages at break of day to weed the dhurra fields. Sown in July, it is harvested in February and March. Eight months are thus required for the cultivation of this cereal in the intense heat of Nubia. For the first three months the growth is extremely rapid, and the stem attains a height of six or seven feet. When at perfection in the rich soil of the Taka country, the plant averages a height of ten feet, the circumference of the stem being about four inches. The crown is a feather very similar to that of the sugar-cane; the blossom falls, and the feather becomes a head of dhurra, weighing about two pounds. Each grain is about the size of hemp-seed. I took the trouble of counting the corns contained in an average- sized head, the result being 4,848. The process of harvesting and threshing is remarkably simple, as the heads are simply detached from the straw and beaten out in piles. The dried straw is a substitute for sticks in forming the walls of the village huts; these are plastered with clay and cow-dung, which form the Arab's lath and plaster.

The millers' work is exclusively the province of the women. No man will condescend to grind the corn. There are no circular hand-mills, as among Oriental nations; but the corn is ground upon a simple flat stone, of cithor gneiss or granite, about two feet in length by fourteen inches in width. The face of this is roughened by beating with a sharp-pointed piece of harder stone, such as quartz or hornblende, and the grain is reduced to flour by great labor and repeated grinding or rubbing with a stone rolling-pin. The flour is mixed with water and allowed to ferment; it is then made into thin pancakes upon an earthenware flat portable hearth. This species of leavened bread is known to the Arabs as the kisra. It is not very palatable, but it is extremely well suited to Arab cookery, as it can be rolled up like a pancake and dipped in the general dish of meat and gravy very conveniently, in the absence of spoons and forks.

On the 14th of July I had concluded my arrangements for the start. There had been some difficulty in procuring camels, but the all-powerful firman was a never-failing talisman, and as the Arabs had declined to let their animals for hire, the Governor despatched a number of soldiers and seized the required number, including their owners. I engaged two wild young Arabs of eighteen and twenty years of age, named Bacheet and Wat Gamma. The latter, being interpreted, signifies "Son of the Moon." This in no way suggests lunacy; but the young Arab had happened to enter this world on the day of the new moon, which was considered to be a particularly fortunate and brilliant omen at his birth. Whether the climax of his good fortune had arrived at the moment he entered my service I know not; but, if so, there was a cloud over his happiness in his subjection to Mahomet, the dragoman, who rejoiced in the opportunity of bullying the two inferiors. Wat Gamma was a quiet, steady, well-conducted lad, who bore oppression mildly ; but the younger, Bucheet, was a fiery, wild young Arab, who, although an excellent boy in his peculiar way, was almost incapable of being tamed and domesticated. I at once perceived that Mahomet would have a determined rebel to control, which I confess I did not regret. Wages were not high in this part of the world--the lads were engaged at one and a half dollars per month and their keep.

Mahomet, who was a great man, suffered from the same complaint to which great men are (in those countries) particularly subject. Wherever he went he was attacked with claimants of relationship. He was overwhelmed with professions of friendship from people who claimed to be connections of some of his family. In fact, if all the ramifications of his race were correctly represented by the claimants of relationship, Mahomet's family tree would have shaded the Nubian desert

We all have our foibles. The strongest fort has its feeble point, as the chain snaps at its weakest link. Family pride was Mahomet's weak link. This was his tender point; and Mahomet, the great and the imperious, yielded to the gentle scratching of his ear if a stranger claimed connection with his ancient lineage. Of course he had no family, with the exception of his wife and two children, whom he had left in Cairo. The lady whom he had honored by admission into the domestic circle of the Mahomets was suffering from a broken arm when we started from Egypt, as she had cooked the dinner badly, and the "gaddah," or large wooden bowl, had been thrown at her by the naturally indignant husband, precisely as he had thrown the axe at one man and the basin at another while in our service. These were little contretemps that could hardly disturb the dignity of so great a man.

Mahomet met several relatives at Cassala. One borrowed money of him; another stole his pipe; the third, who declared that nothing should separate them now that "by the blessing of God" they had met, determined to accompany him through all the difficulties of our expedition, provided that Mahomet would only permit him to serve for love, without wages. I gave Mahomet some little advice upon this point, reminding him that, although the clothes of the party were only worth a few piastres, the spoons and forks were silver; therefore I should hold him responsible for the honesty of his friend. This reflection upon the family gave great offence, and he assured me that Achmet, our quondam acquaintance, was so near a relative that he was-- I assisted him in the genealogical distinction: "Mother's brother's cousin's sister's mother's son? Eh, Mahomet?"

"Yes, sar, that's it!" "Very well, Mahomet; mind he doesn't steal the spoons, and thrash him if he doesn't do his work!" "Yes, sar", replied Mahomet; "he all same like one brother; he one good man; will do his business quietly; if not, master lick him." The new relative not understanding English, was perfectly satisfied with the success of his introduction, and from that moment he became one of the party.

One more addition, and our arrangements were completed: the Governor of Cassala was determined we should not start without a soldier guide to represent the government. Accordingly he gave us a black corporal, so renowned as a sportsman that he went by the name of "El Baggar" (the cow), because of his having killed several of the oryx antelope, known as "El Baggar et Wabash" (cow of the desert).

After sixteen hours' actual marching from Cassala we arrived at the valley of the Atbara. There was an extraordinary change in the appearance of the river between Gozerajup and this spot. There was no longer the vast sandy desert with the river flowing through its sterile course on a level with the surface of the country; but after traversing an apparently perfect flat of forty-five miles of rich alluvial soil, we had suddenly arrived upon the edge of a deep valley, between five and six miles wide, at the bottom of which, about two hundred feet below the general level of the country, flowed the river Atbara. On the opposite side of the valley the same vast table-lands continued to the western horizon.

We commenced the descent toward the river: the valley was a succession of gullies and ravines, of landslips and watercourses. The entire hollow, of miles in width, had evidently been the work of the river. How many ages had the rains and the stream been at work to scoop out from the flat tableland this deep and broad valley? Here was the giant laborer that had shovelled the rich loam upon the delta of Lower Egypt! Upon these vast flats of fertile soil there can be no drainage except through soakage. The deep valley is therefore the receptacle not only for the water that oozes from its sides, but subterranean channels, bursting as land-springs from all parts of the walls of the valley, wash down the more soluble portions of earth, and continually waste away the soil. Landslips occur daily during the rainy season; streams of rich mud pour down the valley's slopes, and as the river flows beneath in a swollen torrent, the friable banks topple down into the stream and dissolve. The Atbara becomes the thickness of peasoup, as its muddy waters steadily perform the duty they have fulfilled from age to age. Thus was the great river at work upon our arrival on its bank at the bottom of the valley. The Arab name, "Bahr el Aswat" (black river) was well bestowed; it was the black mother of Egypt, still carrying to her offspring the nourishment that had first formed the Delta.

At this point of interest the journey had commenced; the deserts were passed; all was fertility and life. Wherever the sources of the Nile might be, the Atbara was the parent of Egypt! This was my first impression, to be proved hereafter.

A violent thunderstorm, with a deluge of rain, broke upon our camp on the banks of the Atbara, fortunately just after the tents were pitched. We thus had an example of the extraordinary effects of the heavy rain in tearing away the soil of the valley. Trifling watercourses were swollen to torrents. Banks of earth became loosened and fell in, and the rush of mud and water upon all sides swept forward into the river with a rapidity which threatened the destruction of the country, could such a tempest endure for a few days. In a couple of hours all was over.

In the evening we crossed with our baggage and people to the opposite side of the ricer, and pitched our tents at the village of Goorashee. In the morning the camels arrived, and once more we were ready to start. Our factotum, El Baggar, had collected a number of baggage-camels and riding dromedaries, or "hygeens". The latter he had brought for approval, as we bad suffered much from the extreme roughness of our late camels. There is the same difference between a good hygeen, or dromedary, and a baggage-camel, as between the thoroughbred and the cart-horse; and it appears absurd in the eyes of the Arabs that a man of any position should ride a baggage-camel. Apart from all ideas of etiquette, the motion of the latter animal is quite sufficient warning. Of all species of fatigue, the back-breaking, monotonous swing of a heavy camel is the worst; and should the rider lose patience and administer a sharp cut with the coorbatch, that induces the creature to break into a trot, the torture of the rack is a pleasant tickling compared to the sensation of having your spine driven by a sledge-hammer from below, half a foot deeper into the skull.

The human frame may be inured to almost anything; thus the Arabs, who have always been accustomed to this kind of exercise, hardly feel the motion, and the portion of the body most subject to pain in riding a rough camel upon two bare pieces of wood for a saddle, becomes naturally adapted for such rough service, as monkeys become hardened from constantly sitting upon rough substances. The children commence almost as soon as they are born, as they must accompany their mothers in their annual migrations; and no sooner can the young Arab sit astride and hold on than he is placed behind his father's saddle, to which he clings, while he bumps upon the bare back of the jolting camel. Nature quickly arranges a horny protection to the nerves, by the thickening of the skin; thus, an Arab's opinion of the action of a riding hygeen should never be accepted without a personal trial. What appears delightful to him may be torture to you, as a strong breeze and a rough sea may be charming to a sailor, but worse than death to a landsman.

I was determined not to accept the camels now offered as hygeens until I had seen them tried. I accordingly ordered our black soldier, El Baggar, to saddle the most easy-actioned animal for my wife; but I wished to see him put it through a variety of paces before she should accept it. The delighted EL

Baggar, who from long practice was as hard as the heel of a boot, disdained a saddle. The animal knelt, was mounted, and off he started at full trot, performing a circle of about fifty yards' diameter as though in a circus. I never saw such an exhibition! "Warranted quiet to ride, of easy action, and fit for a lady!" This had been the character received with the rampant brute, who now, with head and tail erect, went tearing round the circle, screaming and roaring like a wild beast, throwing his forelegs forward and stepping at least three feet high in his trot.

Where was El Baggar? A disjointed looking black figure was sometimes on the back of this easy going camel, sometimes a foot high in the air; arms, head, legs, hands, appeared like a confused mass of dislocation; the woolly hair of this unearthly individual, that had been carefully trained in long stiff narrow curls, precisely similar to the tobacco known as "negro-head," alternately started upright en masse, as though under the influence of electricity, and then fell as suddenly upon his shoulders. Had the dark individual been a "black dose", he or it could not have been more thoroughly shaken. This object, so thoroughly disguised by rapidity of movement, was El Baggar happy, delighted El Baggar! As he came rapidly round toward us flourishing his coorbatch, I called to him, "Is that a nice hygeen for the Sit (lady), EL Baggar? Is it very easy?" He was almost incapable of a reply. "V-e-r-y e-e-a-a-s-y," replied the trustworthy authority, "j-j-j-just the thin-n-n-g for the S-i-i-i-t-t-t." "All right, that will do," I answered, and the jockey pulled up his steed. "Are the other camels better or worse than that?" I asked. "Much worse," replied El Baggar; "the others are rather rough, but this is an easy goer, and will suit the lady well."

It was impossible to hire a good hygeen; an Arab prizes his riding animal too much, and invariably refuses to let it to a stranger, but generally imposes upon him by substituting some lightly-built camel that he thinks will pass muster. I accordingly chose for my wife a steady-going animal from among the baggage-camels, trusting to be able to obtain a hygeen from the great Sheik Abou Sinn, who was encamped upon the road we were about to take along the valley of the Atbara. We left Goorashee on the following day.

CHAPTER III.

The Arabs' exodus-Reception by Abou Sinn-Arabs dressing the hair-Toilet of an Arab woman-The plague of lice-Wives among the Arabs-The Old Testament confirmed

IT was the season of rejoicing. Everybody appeared in good humor. The distended udders of thousands of camels were an assurance of plenty. The burning sun that for nine months had scorched the earth was veiled by passing clouds. The cattle that had panted for water, and whose food was withered straw, were filled with juicy fodder. The camels that had subsisted upon the dried and leafless twigs and branches, now feasted upon the succulent tops of the mimosas. Throngs of women and children mounted upon camels, protected by the peculiar gaudy saddle-hood, ornamented with cowrie- shells, accompanied the march. Thousands of sheep and goats, driven by Arab boys, were straggling in all directions. Baggage-camels, heavily laden with the quaint household goods, blocked up the way. The fine bronzed figures of Arabs, with sword and shield, and white topes, or plaids, guided their milk-white dromedaries through the confused throng with the usual placid dignity of their race, simply passing by with the usual greeting, "Salaam aleikum" (Peace be with you).

It was the Exodus; all were hurrying toward the promised land--"the land flowing with milk and honey", where men and beasts would be secure, not only from the fevers of the south, but from that deadly enemy to camels and cattle, the fly. This terrible insect drove all before it.

If all were right in migrating to the north, it was a logical conclusion that we were wrong in going to the south during the rainy season; however, we now heard from the Arabs that we were within a couple of hours' march from the camp of the great Sheik Achmet Abou Sinn, to whom I had a letter of introduction. At the expiration of about that time we halted, and pitched the tents among some shady mimosas, while I sent Mahomet to Abou Sinn with the letter, and my firman.

I was busily engaged in making sundry necessary arrangements in the tent when Mahomet returned and announced the arrival of the great sheik in person. He was attended by several of his principal people, and as he approached through the bright green mimosas, mounted upon a beautiful snow-white hygeen, I was exceedingly struck with his venerable and dignified appearance. Upon near arrival I went forward to meet him and to assist him from his camel; but his animal knelt immediately at his command, and he dismounted with the ease and agility of a man of twenty.

He was the most magnificent specimen of an Arab that I have ever seen. Although upward of eighty years of age, he was as erect as a lance, and did not appear more than between fifty and sixty. He was of herculean stature, about six feet three inches high, with immensely broad shoulders and chest, a remarkably arched nose, eyes like an eagle's, beneath large, shaggy, but perfectly white eyebrows. A snow-white beard of great thickness descended below the middle of his breast. He wore a large white turban and a white cashmere abbai, or long robe, from the throat to the ankles. As a desert patriarch he was superb--the very perfection of all that the imagination could paint, if we should personify Abraham at the head of his people. This grand old Arab with the greatest politeness insisted upon our immediately accompanying him to his camp, as he could not allow us to remain in his country as strangers. He would hear of no excuses, but at once gave orders to Mahomet to have the baggage repacked and the tents removed, while we were requested to mount two superb white hygeens, with saddle-cloths of blue Persian sheepskins, that he had immediately accoutered when he heard from Mahomet of our miserable camels. The tent was struck, and we joined our venerable host with a line of wild and splendidly-mounted attendants, who followed us toward the sheik's encampment.

Among the retinue of the aged sheik whom we now accompanied, were ten of his sons, some of whom appeared to be quite as old as their father. We had ridden about two miles when we were suddenly met by a crowd of mounted men, armed with the usual swords and shields; many were on horses, others upon hygeens, and all drew up in lines parallel with our approach. These were Abou Sinn's people, who had assembled to give us the honorary welcome as guests of their chief. This etiquette of the Arabs consists in galloping singly at full speed across the line of advance, the rider flourishing the sword over his head, and at the same moment reining up his horse upon its haunches so as to bring it to a sudden halt. This having been performed by about a hundred riders upon both horses and hygeens, they fell into line behind our party, and, thus escorted, we shortly arrived at the Arab encampment. In all countries the warmth of a public welcome appears to be exhibited by noise. The whole neighborhood had congregated to meet us; crowds of women raised the wild, shrill cry that is sounded alike for joy or sorrow; drums were beat; men dashed about with drawn swords and engaged in mimic fight, and in the midst of din and confusion we halted and dismounted. With peculiar grace of manner the old sheik assisted my wife to dismount, and led her to an open shed arranged with angareps (stretchers) covered with Persian carpets and cushions, so as to form a divan. Sherbet, pipes, and coffee were shortly handed to us, and Mahomet, as dragoman, translated the customary interchange of compliments; the sheik assured us that our unexpected arrival among them was "like the blessing of a new moon", the depth of which expression no one can understand who has not experienced life in the desert, where the first faint crescent is greeted with such enthusiasm.

Abou Sinn had arranged to move northward on the following day; we therefore agreed to pass one day in his camp, and to leave the next morning for Sofi, on the Atbara, about seventy-eight miles distant.

From Korosko to this point we had already passed the Bedouins, Bishareens, Hadendowas, Hallongas, until we had entered the Shookeriyahs. On the west of our present position were the Jalyns, and to the south near Sofi were the Dabainas. Many of the tribes claim a right to the title of Bedouins, as descended from that race. The customs of all the Arabs are nearly similar, and the distinction in appearance is confined to a peculiarity in dressing the hair. This is a matter of great importance among both men and women. It would be tedious to describe the minutiae of the various coiffures, but the great desire with all tribes, except the Jalyn, is to have a vast quantity of hair arranged in their own peculiar fashion, and not only smeared, but covered with as much fat as can be made to adhere. Thus, should a man wish to get himself up as a great dandy, he would put at least half a pound of butter or other fat upon his head. This would be worked up with his coarse locks by a friend, until it somewhat resembled a cauliflower. He would then arrange his tope or plaid of thick cotton cloth, and throw one end over his left shoulder, while slung from the same shoulder his circular shield would hang upon his back; suspended by a strap over the right shoulder would hang his long two-edged broadsword.

Fat is the great desideratum of an Arab. His head, as I have described, should be a mass of grease; he rubs his body with oil or other ointment; his clothes, i.e. his one garment or tope, is covered with grease, and internally he swallows as much as he can procure.

The great Sheik Abou Sinn, who is upward of eighty, as upright as a dart, a perfect Hercules, and whose children and grandchildren are like the sand of the sea-shore, has always consumed daily throughout his life two rottolis (pounds) of melted butter. A short time before I left the country he married a new young wife about fourteen years of age. This may be a hint to octogenarians.

The fat most esteemed for dressing the hair is that of the sheep. This undergoes a curious preparation, which renders it similar in appearance to cold cream; upon the raw fat being taken from the animal it is chewed in the mouth by an Arab for about two hours, being frequently taken out for examination during that time, until it has assumed the desired consistency. To prepare sufficient to enable a man to appear in full dress, several persons must be employed in masticating fat at the same time. This species of pomade, when properly made, is perfectly white, and exceedingly light and frothy. It may be imagined that when exposed to a burning sun, the beauty of the head-dress quickly disappears; but the oil then runs down the neck and back, which is considered quite correct, especially when the tope becomes thoroughly greased. The man is then perfectly anointed. We had seen an amusing example of this when on the march from Berber to Gozerajup. The Turk, Hadji Achmet, had pressed into our service, as a guide for a few miles, a dandy who had just been arranged as a cauliflower, with at least half a pound of white fat upon his head. As we were travelling upward of four miles an hour in an intense heat, during which he was obliged to run, the fat ran quicker than he did, and at the end of a couple of hours both the dandy and his pomade were exhausted. The poor fellow had to return to his friends with the total loss of personal appearance and half a pound of butter.

Not only are the Arabs particular in their pomade, but great attention is bestowed upon perfumery, especially by the women. Various perfumes are brought from Cairo by the travelling native merchants, among which those most in demand are oil of roses, oil of sandal-wood, an essence from the blossom of a species of mimosa, essence of musk, and the oil of cloves. The women have a peculiar method of scenting their bodies and clothes by an operation that is considered to be one of the necessaries of life, and which is repeated at regular intervals. In the floor of the tent, or hut, as it may chance to be, a small hole is excavated sufficiently large to contain a common-sized champagne bottle. A fire of charcoal, or of simply glowing embers, is made within the hole, into which the woman about to be scented throws a handful of various drugs. She then takes off the cloth or tope which forms her dress, and crouches naked over the fumes, while she arranges her robe to fall as a mantle from her neck to the ground like a tent. When this arrangement is concluded she is perfectly happy, as none of the precious fumes can escape, all being retained beneath the robe, precisely as if she wore a crinoline with an incense-burner beneath it, which would be a far more simple way of performing the operation. She now begins to perspire freely in the hot-air bath, and the pores of the skin being thus opened and moist, the volatile oil from the smoke of the burning perfumes is immediately absorbed.

By the time that the fire has expired the scenting process is completed, and both her person and robe are redolent of incense, with which they are so thoroughly impregnated that I have frequently smelt a party of women strongly at full a hundred yards' distance, when the wind has been blowing from their direction.

The Arab women do not indulge in fashions. Strictly conservative in their manners and customs, they never imitate, but they simply vie with each other in the superlativeness of their own style; thus the dressing of the hair is a most elaborate affair, which occupies a considerable portion of their time. It is quite impossible for an Arab woman to arrange her own hair; she therefore employs an assistant, who, if clever in the art, will generally occupy about three days before the operation is concluded. First, the hair must be combed with a long skewer-like pin; then, when well divided, it becomes possible to use an exceedingly coarse wooden comb. When the hair is reduced to reasonable order by the latter process, a vigorous hunt takes place, which occupies about an hour, according to the amount of game preserved. The sport concluded, the hair is rubbed with a mixture of oil of roses, myrrh, and sandal-wood dust mixed with a powder of cloves and cassia. When well greased and rendered somewhat stiff by the solids thus introduced, it is plaited into at least two hundred fine plaits; each of these plaits is then smeared with a mixture of sandal-wood dust and either gum water or paste of dhurra flour. On the last day of the operation, each tiny plait is carefully opened by the long hairpin or skewer, and the head is ravissante. Scented and frizzled in this manner with a well-greased tope or robe, the Arab lady's toilet is complete. Her head is then a little larger than the largest sized English mop, and her perfume is something between the aroma of a perfumer's shop and the monkey-house at the Zoological Gardens. This is considered "very killing," and I have been quite of that opinion when a crowd of women have visited my wife in our tent, with the thermometer at 95 degrees C, and have kindly consented to allow me to remain as one of the party.

It is hardly necessary to add that the operation of hairdressing is not often performed, but that the effect is permanent for about a week, during which time the game becomes so excessively lively that the creatures require stirring up with the long hairpin or skewer whenever too unruly. This appears to be constantly necessary from the vigorous employment of the ruling sceptre during conversation. A levee of Arab women in the tent was therefore a disagreeable invasion, as we dreaded the fugitives; fortunately, they appeared to cling to the followers of Mahomet in preference to Christians.

The plague of lice brought upon the Egyptians by Moses has certainly adhered to the country ever since, if "lice" is the proper translation of the Hebrew word in the Old Testament. It is my own opinion that the insects thus inflicted upon the population were not lice, but ticks. Exod. 8:16: "The dust became lice throughout all Egypt;" again, Exod. 8:17: "Smote dust... it became lice in man and beast." Now the louse that infests the human body and hair has no connection whatever with "dust," and if subject to a few hours' exposure to the dry heat of the burning sand, it would shrivel and die. But the tick is an inhabitant of the dust, a dry horny insect without any apparent moisture in its composition; it lives in hot sand and dust, where it cannot possibly obtain nourishment, until some wretched animal lies down upon the spot, when it becomes covered with these horrible vermin. I have frequently seen dry desert places so infested with ticks that the ground was perfectly alive with them, and it would have been impossible to rest on the earth.

In such spots, the passage in Exodus has frequently occurred to me as bearing reference to these vermin, which are the greatest enemies to man and beast. It is well known that, from the size of a grain of sand in their natural state, they will distend to the size of a hazelnut after having preyed for some days upon the blood of an animal. The Arabs are invariably infested with lice, not only in their hair, but upon their bodies and clothes; even the small charms or spells worn upon the arm in neatly-sewn leathern packets are full of these vermin. Such spells are generally verses copied from the Koran by the Faky, or priest, who receives some small gratuity in exchange. The men wear several such talismans upon the arm above the elbow, but the women wear a large bunch of charms, as a sort of chatelaine, suspended beneath their clothes around the waist.

Although the tope or robe, loosely but gracefully arranged around the body, appears to be the whole of the costume, the women wear beneath this garment a thin blue cotton cloth tightly bound round the loins, which descends to a little above the knee; beneath this, next to the skin, is the last garment, the rahat. The latter is the only clothing of young girls, and may be either perfectly simple or adorned with beads and cowrie shells according to the fancy of the wearer. It is perfectly effective as a dress, and admirably adapted to the climate.

The rahat is a fringe of fine dark brown or reddish twine, fastened to a belt, and worn round the waist. On either side are two long tassels, that are generally ornamented with beads or cowries, and dangle nearly to the ankles, while the rahat itself should descend to a little above the knee, or be rather shorter than a Highland kilt. Nothing can be prettier or more simple than this dress, which, although short, is of such thickly hanging fringe that it perfectly answers the purpose for which it is intended.

Many of the Arab girls are remarkably good-looking, with fine figures until they become mothers. They generally marry at the age of thirteen or fourteen, but frequently at twelve or even earlier. Until married, the rahat is their sole garment. Throughout the Arab tribes of Upper Egypt, chastity is a necessity, as an operation is performed at the early age of from three to five years that thoroughly protects all females and which renders them physically proof against incontinency.

There is but little love-making among the Arabs. The affair of matrimony usually commences by a present to the father of the girl, which, if accepted, is followed by a similar advance to the girl herself, and the arrangement is completed. All the friends of both parties are called together for the wedding; pistols and guns are fired off, if possessed. There is much feasting, and the unfortunate bridegroom undergoes the ordeal of whipping by the relatives of his bride, in order to test his courage. Sometimes this punishment is exceedingly severe, being inflicted with the coorbatch or whip of hippopotamus hide, which is cracked vigorously about his ribs and back. If the happy husband wishes to be considered a man worth having, he must receive the chastisement with an expression of enjoyment; in which case the crowds of women again raise their thrilling cry in admiration. After the rejoicings of the day are over, the bride is led in the evening to the residence of her husband, while a beating of drums and strumming of guitars (rhababas) are kept up for some hours during the night, with the usual discordant singing.

There is no divorce court among the Arabs. They are not sufficiently advanced in civilization to accept a pecuniary fine as the price of a wife's dishonor; but a stroke of the husband's sword or a stab with the knife is generally the ready remedy for infidelity. Although strict Mahometans, the women are never veiled; neither do they adopt the excessive reserve assumed by the Turks and Egyptians. The Arab women are generally idle, and one of the conditions of accepting a suitor is that a female slave is to be provided for the special use of the wife. No Arab woman will engage herself as a domestic servant; thus, so long as their present customs shall remain unchanged, slaves are creatures of necessity. Although the law of Mahomet limits the number of wives for each man to four at one time, the Arab women do not appear to restrict their husbands to this allowance, and the slaves of the establishment occupy the position of concubines.

The Arabs adhere strictly to their ancient customs, independently of the comparatively recent laws established by Mahomet. Thus, concubinage is not considered a breach of morality; neither is it regarded by the legitimate wives with jealousy. They attach great importance to the laws of Moses and to the customs of their forefathers; neither can they understand the reason for a change of habit in any respect where necessity has not suggested the reform. The Arabs are creatures of necessity; their nomadic life is compulsory, as the existence of their flocks and herds depends upon the pasturage. Thus, with the change of seasons they must change their localities, according to the presence of fodder for their cattle. Driven to and fro by the accidents of climate, the Arab has been compelled to become a wanderer; and precisely as the wild beasts of the country are driven from place to place either by the arrival of the fly, the lack of pasturage, or by the want of water, even so must the flocks of the Arab obey the law of necessity, in a country where the burning sun and total absence of rain for nine months of the year convert the green pastures into a sandy desert.

The Arab cannot halt on one spot longer than the pasturage will support his flocks; therefore his necessity is food for his beasts. The object of his life being fodder, he must wander in search of the ever-changing supply. His wants must be few, as the constant changes of encampment necessitate the transport of all his household goods; thus he reduces to a minimum the domestic furniture and utensils. No desires for strange and fresh objects excite his mind to improvement, or alter his original habits; he must limit his impedimenta, not increase them. Thus with a few necessary articles he is contented. Mats for his tent, ropes manufactured with the hair of his goats and camels, pots for carrying fat, water-jars and earthenware pots or gourd-shells for containing milk, leather water-skins for the desert, and sheep-skin bags for his clothes--these are the requirements of the Arabs. Their patterns have never changed, but the water-jar of to-day is of the same form as that carried to the well by the women of thousands of years ago. The conversation of the Arabs is in the exact style of the Old Testament. The name of God is coupled with every trifling incident in life, and they believe in the continual action of divine special interference. Should a famine afflict the country, it is expressed in the stern language of the bible--"The Lord has sent a grievous famine upon the land;" or, "The Lord called for a famine, and it came upon the land." Should their cattle fall sick, it is considered to be an affliction by divine command; or should the flocks prosper and multiply particularly well during one season, the prosperity is attributed to special interference. Nothing can happen in the usual routine of daily life without a direct connection with the hand of God, according to the Arab's belief.

This striking similarity to the descriptions of the Old Testament is exceedingly interesting to a traveller when residing among these curious and original people. With the Bible in one hand, and these unchanged tribes before the eyes, there is a thrilling illustration of the sacred record; the past becomes the present; the veil of three thousand years is raised, and the living picture is a witness to the exactness of the historical description. At the same time there is a light thrown upon many obscure passages in the Old Testament by a knowledge of the present customs and figures of speech of the Arabs, which are precisely those that were practised at the periods described. I do not attempt to enter upon a theological treatise, therefore it is unnecessary to allude specially to these particular points. The sudden and desolating arrival of a flight of locusts, the plague, or any other unforeseen calamity, is attributed to the anger of God, and is believed to be an infliction of punishment upon the people thus visited, precisely as the plagues of Egypt were specially inflicted upon Pharaoh and the Egyptians.

Should the present history of the country be written by an Arab scribe, the style of the description would be purely that of the Old Testament; and the various calamities or the good fortunes that have in the course of nature befallen both the tribes and individuals would be recounted either as special visitations of divine wrath or blessings for good deeds performed. If in a dream a particular course of action is suggested, the Arab believes that God has spoken and directed him. The Arab scribe or historian would describe the event as the "voice of the Lord" ("kallam el Allah"), having spoken unto the person; or, that God appeared to him in a dream and "said," etc. Thus much allowance would be necessary on the part of a European reader for the figurative ideas and expressions of the people. As the Arabs are unchanged, the theological opinions which they now hold are the same as those which prevailed in remote ages, with the simple addition of their belief in Mahomet as the Prophet.

CHAPTER IV.

On the Abyssinian border. A new school of medicine--Sacred shrines and epidemics.

We left the camp of Abou Sinn on the morning of July 25th, and in a few rapid marches arrived at Tomat, a lovely spot at the junction of the Atbara with the Settite.

The Settite is the river par excellence, as it is the principal stream of Abyssinia, in which country it bears the name of "Tacazzy." Above the junction the Athara does not exceed two hundred yards in width. Both rivers have scooped out deep and broad valleys throughout their course. This fact confirmed my first impression that the supply of soil had been brought down by the Atbara to the Nile. The country on the opposite or eastern bank of the Atbara is contested ground. In reality it forms the western frontier of

Abyssinia, of which the Atbara River is the boundary; but since the annexation of the Nubian provinces to Egypt there has been no safety for life or property upon the line of frontier; thus a large tract of country actually forming a portion of Abyssinia is uninhabited.

Upon our arrival at Sofi we were welcomed by the sheik, and by a German, Florian, who was delighted to see Europeans. He was a sallow, sickly-looking man, who with a large bony frame had been reduced from constant hard work and frequent sickness to little but skin and sinew. He was a mason, who had left Germany with the Austrian mission to Khartoum, but finding the work too laborious in such a climate, he and a friend, who was a carpenter, had declared for independence, and they had left the mission. They were both enterprising fellows, and sportsmen; therefore they had purchased rifles and ammunition, and had commenced life as hunters. At the same time they employed their leisure hours in earning money by the work of their hands in various ways.

I determined to arrange our winter quarters at Sofi for three months' stay, during which I should have ample time to gain information and complete arrangements for the future. I accordingly succeeded in purchasing a remarkably neat house for ten piastres (two shillings). The architecture was of an ancient style, from the original design of a pill-box surmounted by a candle extinguisher. I purchased two additional huts, which were erected at the back of our mansion, one as the kitchen, the other as the servants' hall.

In the course of a week we had as pretty a camp as Robinson Crusoe himself could have coveted. We had a view of about five miles in extent along the valley of the Atbara, and it was my daily amusement to scan with my telescope the uninhabited country upon the opposite side of the river and watch the wild animals as they grazed in perfect security. We were thoroughly happy at Sofi. There was a delightful calm and a sense of rest, a total estrangement from the cares of the world, and an enchanting contrast in the soft green verdure of the landscape before us, to the many hundred weary miles of burning desert through which we had toiled from Lower Egypt.

Time glided away smoothly until the fever invaded our camp. Florian became seriously ill. My wife was prostrated by a severe attack of gastric fever, which for nine days rendered her recovery almost hopeless. Then came the plague of boils, and soon after a species of intolerable itch, called the coorash. I adopted for this latter a specific I had found successful with the mange in dogs, namely, gunpowder, with one fourth sulphur added, made into a soft paste with water, and then formed into an ointment with fat. It worked like a charm with the coorash.

Faith is the drug that is supposed to cure the Arab; whatever his complaint may be, he applies to his Faky or priest. This minister is not troubled with a confusion of book-learning, neither are the shelves of his library bending beneath weighty treatises upon the various maladies of human nature; but he possesses the key to all learning, the talisman that will apply to all cases, in that one holy book, the Koran. This is his complete pharmacopoeia: his medicine chest, combining purgatives, blisters, sudorifies, styptics, narcotics, emetics, and all that the most profound M.D. could prescribe. With this "multum in parvo" stock-in-trade the Faky receives his patients. No. 1 arrives, a barren woman who requests some medicine that will promote the blessing of childbirth. No. 2, a man who was strong in his youth, but from excessive dissipation has become useless. No. 3, a man deformed from his birth, who wishes to become straight as other men. No. 4, a blind child. No. 5, a dying old woman, carried on a litter; and sundry other impossible cases, with others of a more simple character.

The Faky produces his book, the holy Koran, and with a pen formed of a reed he proceeds to write a prescription--not to be made up by an apothecary, as such dangerous people do not exist; but the prescription itself is to be SWALLOWED! Upon a smooth board, like a slate, he rubs sufficient lime to produce a perfectly white surface; upon this he writes in large characters, with thick glutinous ink, a verse or verses from the Koran that he considers applicable to the case; this completed, he washes off the holy quotation, and converts it into a potation by the addition of a little water; this is swallowed in perfect faith by the patient, who in return pays a fee according to the demand of the Faky.

As few people can read or write, there is an air of mystery in the art of writing which much enhances the value of a scrap of paper upon which is written a verse from the Koran. A few piastres are willingly expended in the purchase of such talismans, which are carefully and very neatly sewn into small envelopes of leather, and are worn by all people, being handed down from father to son.

The Arabs are especially fond of relics; thus, upon the return from a pilgrimage to Mecca, the "hadji" or pilgrim is certain to have purchased from some religious Faky of the sacred shrine either a few square inches of cloth, or some such trifle, that belonged to the prophet Mahomet. This is exhibited to his friends and strangers as a wonderful spell against some particular malady, and it is handed about and received with extreme reverence by the assembled crowd. I once formed one of a circle when a pilgrim returned to his native village. We sat in a considerable number upon the ground, while he drew from his bosom a leather envelope, suspended from his neck, from which he produced a piece of extremely greasy woollen cloth, about three inches square, the original color of which it would have been impossible to guess. This was a piece of Mahomet's garment, but what portion he could not say. The pilgrim had paid largely for this blessed relic, and it was passed round our circle from hand to hand, after having first been kissed by the proprietor, who raised it to the crown of his head, which he touched with the cloth, and then wiped both his eyes. Each person who received it went through a similar performance, and as ophthalmia and other diseases of the eyes were extremely prevalent, several of the party had eyes that had not the brightness of the gazelle's; nevertheless, these were supposed to become brighter after having been wiped by the holy cloth. How many eyes this same piece of cloth had wiped, it would be impossible to say, but such facts are sufficient to prove the danger of holy relics, that are inoculators of all manner of contagious diseases.

I believe in holy shrines as the pest spots of the world. We generally have experienced in Western Europe that all violent epidemics arrive from the East. The great breadth of the Atlantic boundary would naturally protect us from the West, but infectious disorders, such as plague, cholera, small-pox, etc., may be generally tracked throughout their gradations from their original nests. Those nests are in the East, where the heat of the climate acting upon the filth of semi-savage communities engenders pestilence.

The holy places of both Christians and Mahometans are the receptacles for the masses of people of all nations and classes who have arrived from all points of the compass. The greater number of such people are of poor estate, and many have toiled on foot from immense distances, suffering from hunger and fatigue, and bringing with them not only the diseases of their own remote counties, but arriving in that weak state that courts the attack of any epidemic. Thus crowded together, with a scarcity of provisions, a want of water, and no possibility of cleanliness, with clothes that have been unwashed for weeks or months, in a camp of dirty pilgrims, without any attempt at drainage, an accumulation of filth takes place that generates either cholera or typhus; the latter, in its most malignant form, appears as the dreaded "plague." Should such an epidemic attack the mass of pilgrims debilitated by the want of nourishing food, and exhausted by their fatiguing march, it runs riot like a fire among combustibles, and the loss of life is terrific. The survivors radiate from this common centre, upon their return to their respective homes, to which they carry the seeds of the pestilence to germinate upon new soils in different countries. Doubtless the clothes of the dead furnish materials for innumerable holy relics as vestiges of the wardrobe of the Prophet. These are disseminated by the pilgrims throughout all countries, pregnant with disease; and, being brought into personal contact with hosts of true believers, Pandora's box could not be more fatal.

Not only are relics upon a pocket scale conveyed by pilgrims and reverenced by the Arabs, but the body of any Faky who in lifetime was considered unusually holy is brought from a great distance to be interred in some particular spot. In countries where a tree is a rarity, a plank for a coffin is unknown; thus the reverend Faky, who may have died of typhus, is wrapped in cloths and packed in a mat. In this form he is transported, perhaps some hundred miles, slung upon a camel, with the thermometer above 130 degrees Fah. in the sun, and he is conveyed to the village that is so fortunate as to be honored with his remains. It may be readily imagined that with a favorable wind the inhabitants are warned of his approach some time before his arrival.

Happily, long before we arrived at Sofi, the village had been blessed by the death of a celebrated Faky, a holy man who would have been described as a second Isaiah were the annals of the country duly chronicled. This great "man of God," as he was termed, had departed this life at a village on the borders of the Nile, about eight days' hard camel-journey from Sofi; but from some assumed right, mingled no doubt with jobbery, the inhabitants of Sofi had laid claim to his body, and he had arrived upon a camel horizontally, and had been buried about fifty yards from the site of our camp. His grave was beneath a clump of mimosas that shaded the spot, and formed the most prominent object in the foreground of our landscape. Thither every Friday the women of the village congregated, with offerings of a few handfuls of dhurra in small gourd-shells, which they laid upon the grave, while they ATE THE HOLY EARTH in small pinches, which they scraped like rabbits, from a hole they had burrowed toward the venerated corpse. This hole was about two feet deep from continual scratching, and must have been very near the Faky.

Although thus reverent in their worship, the Arab's religion is a sort of adjustable one. The wild boar, for instance, is invariably eaten by the Arab hunters, although in direct opposition to the rules of the Koran. I once asked them what their Faky would say if he were aware of such a transgression. "Oh!" they replied, "we have already asked his permission, as we are sometimes severely pressed for food in the jungles. He says, `If you have the KORAN in your hand and NO PIG, you are forbidden to eat pork; but if you have the PIG in your hand and NO KORAN, you had better eat what God has given you.'"

CHAPTER V.

A primitive craft--Stalking the giraffes--My first giraffes--Rare sport with the finny tribe--Thieving elephants.

For many days, while at Sofi, we saw large herds of giraffes and antelopes on the opposite side of the river, about two miles distant. On September 2d a herd of twenty-eight giraffes tempted me at all hazards to cross the river. So we prepared an impromptu raft. My angarep (bedstead) was quickly inverted. Six water-skins were inflated, and lashed, three on either side. A shallow packing- case, lined with tin, containing my gun, was fastened in the centre of the angarep, and two towlines were attached to the front part of the raft, by which swimmers were to draw it across the river. Two men were to hang on behind, and, if possible, keep it straight in the rapid current. After some difficulty we arrived at the opposite bank, and scrambled through thick bushes, upon our hands and knees, to the summit.

For about two miles' breadth on this side of the river the valley was rough broken ground, full of gullies and ravines sixty or seventy feet deep, beds of torrents, bare sandstone rocks, bushy crags, fine grassy knolls, and long strips of mimosa covert, forming a most perfect locality for shooting.

I had observed by the telescope that the giraffes were standing as usual upon an elevated position, from whence they could keep a good lookout. I knew it would be useless to ascend the slope directly, as their long necks give these animals an advantage similar to that of the man at the masthead; therefore, although we had the wind in our favor, we should have been observed. I accordingly determined to make a great circuit of about five miles, and thus to approach them from above, with the advantage of the broken ground for stalking. It was the perfection of uneven country. By clambering up broken cliffs, wading shoulder-deep through muddy gullies, sliding down the steep ravines, and winding through narrow bottoms of high grass and mimosas for about two hours, we at length arrived at the point of the high table-land upon the verge of which I had first noticed the giraffes with the telescope. Almost immediately I distinguished the tall neck of one of these splendid animals about half a mile distant upon my left, a little below the table-land; it was feeding on the bushes, and I quickly discovered several others near the leader of the herd. I was not far enough advanced in the circuit that I had intended to bring me exactly above them, therefore I turned sharp to my right, intending to make a short half circle, and to arrive on the leeward side of the herd, as I was now to windward. This I fortunately completed, but I had marked a thick bush as my point of cover, and upon arrival I found that the herd had fed down wind, and that I was within two hundred yards of the great bull sentinel that, having moved from his former position, was now standing directly before me.

I lay down quietly behind the bush with my two followers, and anxiously watched the great leader, momentarily expecting that it would get my wind. It was shortly joined by two others, and I perceived the heads of several giraffes lower down the incline, that were now feeding on their way to the higher ground. The seroot fly was teasing them, and I remarked that several birds were fluttering about their heads, sometimes perching upon their noses and catching the fly that attacked their nostrils, while the giraffes appeared relieved by their attentions. These birds were of a peculiar species that attacks the domestic animals, and not only relieves them of vermin, but eats into the flesh and establishes dangerous sores. A puff of wind now gently fanned the back of my neck; it was cool and delightful, but no sooner did I feel the refreshing breeze than I knew it would convey our scent directly to the giraffes. A few seconds afterward the three grand obelisks threw their heads still higher in the air, and fixing their great black eyes upon the spot from which the warning came, they remained as motionless as though carved from stone. From their great height they could see over the bush behind which we were lying at some paces distant, and although I do not think they could distinguish us to be men, they could see enough to convince them of hidden enemies.

The attitude of fixed attention and surprise of the three giraffes was sufficient warning for the rest of the herd, who immediately filed up from the lower ground, and joined their comrades. All now halted and gazed steadfastly in our direction, forming a superb tableau, their beautiful mottled skins glancing like the summer coat of a thoroughbred horse, the orange-colored statues standing out in high relief from a background of dark-green mimosas.

This beautiful picture soon changed. I knew that my chance of a close shot was hopeless, as they would presently make a rush and be off; thus I determined to get the first start. I had previously studied the ground, and I concluded that they would push forward at right angles with my position, as they had thus ascended the hill, and that, on reaching the higher ground, they would turn to the right, in order to reach an immense tract of high grass, as level as a billiard-table, from which no danger could approach them unobserved.

I accordingly with a gentle movement of my hand directed my people to follow me, and I made a sudden rush forward at full speed. Off went the herd, shambling along at a tremendous pace, whisking their long tails above their hind quarters, and, taking exactly the direction I had anticipated, they offered me a shoulder shot at a little within two hundred yards' distance. Unfortunately, I fell into a deep hole concealed by the high grass, and by the time that I resumed the hunt they had increased their distance; but I observed the leader turned sharply to the right, through some low mimosa bush, to make directly for the open table-land. I made a short cut obliquely at my best speed, and only halted when I saw that I should lose ground by altering my position. Stopping short, I was exactly opposite the herd as they filed by me at right angles in full speed, within about a hundred and eighty yards. I had my old Ceylon No. 10 double rifle, and I took a steady shot at a large dark-colored bull. The satisfactory sound of the ball upon his hide was followed almost immediately by his blundering forward for about twenty yards and falling heavily in the low bush. I heard the crack of the ball of my left-hand barrel upon another fine beast, but no effects followed. Bacheet quickly gave me the single two-ounce Manton rifle, and I singled out a fine dark-colored bull, who fell on his knees to the shot, but, recovering, hobbled off disabled, apart from the herd, with a foreleg broken just below the shoulder. Reloading immediately, I ran up to the spot, where I found my first giraffe lying dead, with the ball clean through both shoulders. The second was standing about one hundred paces distant. Upon my approach he attempted to move, but immediately fell, and was despatched by my eager Arabs. I followed the herd for about a mile to no purpose, through deep clammy ground and high grass, and I returned to our game.

These were my first giraffes, and I admired them as they lay before me with a hunter's pride and satisfaction, but mingled with a feeling of pity for such beautiful and utterly helpless creatures. The giraffe, although from sixteen to twenty feet in height, is perfectly defenceless, and can only trust to the swiftness of its pace and the extraordinary power of vision, for its means of protection. The eye of this animal is the most beautiful exaggeration of that of the gazelle, while the color of the reddish-orange hide, mottled with darker spots, changes the tints of the skin with the differing rays of light, according to the muscular movement of the body. No one who has merely seen the giraffe in a cold climate can form the least idea of its beauty in its native land.

Life at Sofi was becoming sadly monotonous, and I determined to move my party across the river to camp on the uninhabited side. The rains had almost ceased, so we should be able to live in a tent by night, and to form a shady nook beneath some mimosas by day. On the 15th of September the entire male population of Sofi turned out to assist us across the river. I had arranged a raft by attaching eight inflated skins to the bedstead, upon which I lashed our large circular sponging bath. Four hippopotami hunters were harnessed as tug steamers. By evening all our party, with the baggage, had effected the crossing without accident--all but Achmet, Mahomet's mother's brother's cousin's sister's mother's son, who took advantage of his near relative, when the latter was in the middle of the stream, and ran off with most of his personal effects.

The life at our new camp was charmingly independent. We were upon Abyssinian territory, but as the country was uninhabited we considered it as our own. Our camp was near the mouth of a small stream, the Till, tributary to the Atbara, which afforded some excellent sport in fishing. Choosing one day a fish of about half a pound for bait, I dropped this in the river about twenty yards beyond the mouth of the Till, and allowed it to swim naturally down the stream so as to pass across the Till junction, and descend the deep channel between the rocks. For about ten minutes I had no run. I had twice tried the same water without success; nothing would admire my charming bait; when, just as it had reached the favorite turning-point at the extremity of a rock, away dashed the line, with the tremendous rush that follows the attack of a heavy fish. Trusting to the soundness of my tackle, I struck hard and fixed my new acquaintance thoroughly, but off he dashed down the stream for about fifty yards at one rush, making for a narrow channel between two rocks, through which the stream ran like a mill-race. Should he pass this channel, I knew he would cut the line across the rock; therefore, giving him the butt, I held him by main force, and by the great swirl in the water I saw that I was bringing him to the surface; but just as I expected to see him, my float having already appeared, away he darted in another direction, taking sixty or seventy yards of line without a check. I at once observed that he must pass a shallow sandbank favorable for landing a heavy fish; I therefore checked him as he reached this spot, and I followed him down the bank, reeling up line as I ran parallel with his course. Now came the tug of war! I knew my hooks were good and the line sound, therefore I was determined not to let him escape beyond the favorable ground; and I put upon him a strain that, after much struggling, brought to the surface a great shovel-head, followed by a pair of broad silvery sides, as I led him gradually into shallow water. Bacheet now cleverly secured him by the gills, and dragged him in triumph to the shore. This was a splendid bayard, of at least forty pounds' weight.

I laid my prize upon some green reeds, and covered it carefully with the same cool material. I then replaced my bait by a lively fish, and once more tried the river. In a very short time I had another run, and landed a small fish of about nine pounds, of the same species. Not wishing to catch fish of that size, I put on a large bait, and threw it about forty yards into the river, well up the stream, and allowed the float to sweep the water in a half circle, thus taking the chance of different distances from the shore. For about half an hour nothing moved. I was just preparing to alter my position, when out rushed my line, and, striking hard, I believed I fixed the old gentleman himself, for I had no control over him whatever. Holding him was out of the question; the line flew through my hands, cutting them till the blood flowed, and I was obliged to let the fish take his own way. This he did for about eighty yards, when he suddenly stopped. This unexpected halt was a great calamity, for the reel overran itself, having no checkwheel, and the slack bends of the line caught the handle just as he again rushed forward, and with a jerk that nearly pulled the rod from my hands he was gone! I found one of my large hooks broken short off. The fish was a monster!

After this bad luck I had no run until the evening, when, putting on a large bait, and fishing at the tail of a rock between the stream and still water, I once more had a fine rush, and hooked a big one. There were no rocks down stream, all was fair play and clear water, and away he went at racing pace straight for the middle of the river. To check the pace I grasped the line with the stuff of my loose trousers, and pressed it between my fingers so as to act as a brake and compel him to labor for every yard; but he pulled like a horse, and nearly cut through the thick cotton cloth, making straight running for at least a hundred yards without a halt. I now put so severe a strain upon him that my strong bamboo bent nearly double, and the fish presently so far yielded to the pressure that I could enforce his running in half circles instead of straight away. I kept gaining line until I at length led him into a shallow bay, and after a great fight Bacheet embraced him by falling upon him and clutching the monster with hands and knees; he then tugged to the shore a magnificent fish of upward of sixty pounds. For about twenty minutes lie had fought against such a strain as I had never before used upon a fish; but I had now adopted hooks of such a large size and thickness that it was hardly possible for them to break, unless snapped by a crocodile. My reel was so loosened from the rod, that had the struggle lasted a few minutes longer I must have been vanquished. This fish measured three feet eight inches to the root of the tail, and two feet three inches in girth of shoulders ; the head measured one foot ten inches in circumference. It was of the same species as those I had already caught.

Over a month was passed at our camp, Ehetilla, as we called it. The time passed in hunting, fishing, and observing the country, but it was for the most part uneventful. In the end of October we removed to a village called Wat el

Negur, nine miles south-east of Ehetilla, still on the bank of the Atbara.

Our arrival was welcomed with enthusiasm. The Arabs here had extensive plantations of sesame, dhurra, and cotton, and the nights were spent in watching them, to scare away the elephants, which, with extreme cunning, invaded the fields of dhurra at different points every night, and retreated before morning to the thick, thorny jungles of the Settite. The Arabs were without firearms, and the celebrated aggageers or sword-hunters were useless, as the elephants appeared only at night, and were far too cunning to give them a chance. I was importuned to drive away the elephants, and one evening, about nine o'clock, I arrived at the plantations with three men carrying spare guns. We had not been half an hour in the dhurra fields before we met a couple of Arab watchers, who informed us that a herd of elephants was already in the plantation; we accordingly followed our guides. In about a quarter of an hour we distinctly heard the cracking of the dhurra stems, as the elephants browsed and trampled them beneath their feet.

Taking the proper position of the wind, I led our party cautiously in the direction of the sound, and in about five minutes I came in view of the slate-colored and dusky forms of the herd. The moon was bright, and I counted nine elephants; they had trampled a space of about fifty yards square into a barren level, and they were now slowly moving forward, feeding as they went. One elephant, unfortunately, was separated from the herd, and was about forty yards in the rear; this fellow I was afraid would render our approach difficult. Cautioning my men, especially Bacheet, to keep close to me with the spare rifles, I crept along the alleys formed by the tall rows of dhurra, and after carefully stalking against the wind, I felt sure that it would be necessary to kill the single elephant before I should be able to attack the herd. Accordingly I crept nearer and nearer, well concealed in the favorable crop of high and sheltering stems, until I was within fifteen yards of the hindmost animal. As I had never shot one of the African species, I was determined to follow the Ceylon plan, and get as near as possible; therefore I continued to creep from row to row of dhurra, until I at length stood at the very tail of the elephant in the next row. I could easily have touched it with my rifle, but just at this moment it either obtained my wind or it heard the rustle of the men. It quickly turned its head half round toward me; in the same instant I took the temple-shot, and by the flash of the rifle I saw that it fell. Jumping forward past the huge body, I fired the left-hand barrel at an elephant that had advanced from the herd; it fell immediately! Now came the moment for a grand rush, as they stumbled in confusion over the last fallen elephant, and jammed together in a dense mass with their immense ears outspread, forming a picture of intense astonishment! Where were my spare guns? Here was an excellent opportunity to run in and floor them right and left!

Not a man was in sight! Everybody had bolted, and I stood in advance of the dead elephant calling for my guns in vain. At length one of my fellows came up, but it was too late. The fallen elephant in the herd had risen from the ground, and they had all hustled off at a great pace, and were gone. I had only bagged one elephant. Where was the valiant Bacheet--the would-be Nimrod, who for the last three months had been fretting in inactivity, and longing for the moment of action, when he had promised to be my trusty gun-bearer? He was the last man to appear, and he only ventured from his hiding-place in the high dhurra when assured of the elephants' retreat. I was obliged to admonish the whole party by a little physical treatment, and the gallant Bacheet returned with us to the village, crestfallen and completely subdued. On the following day not a vestige remained of the elephant, except the offal; the Arabs had not only cut off the flesh, but they had hacked the skull and the bones in pieces, and carried them off to boil down for soup.

CHAPTER VI.

Preparations for advance--Mek Nimmur makes a foray--The Hamran elephant-hunters--In the haunts of the elephant--A desperate charge.

The time was approaching when the grass throughout the country would be sufficiently dry to be fired. We accordingly prepared for our expedition; but it was first necessary for me to go to Katariff, sixty miles distant, to engage men, and to procure a slave in place of old Masara, whose owner would not trust her in the wild region we were about to visit.

I engaged six strong Tokrooris for five months, and purchased a slave woman for thirty-five dollars. The name of the woman was Barrake. She was about twenty-two years of age, brown in complexion, fat and strong, rather tall, and altogether she was a fine, powerful-looking woman, but decidedly not pretty. Her hair was elaborately dressed in hundreds of long narrow curls, so thickly smeared with castor oil that the grease had covered her naked shoulders. In addition to this, as she had been recently under the hands of the hairdresser, there was an amount of fat and other nastiness upon her head that gave her the appearance of being nearly gray.

Through the medium of Mahomet I explained to her that she was no longer a slave, as I had purchased her freedom; that she would not even be compelled to remain with us, but she could do as she thought proper; that both her mistress and I should be exceedingly kind to her, and we would subsequently find her a good situation in Cairo; in the mean time she would receive good clothes and wages. This, Mahomet, much against his will, was obliged to translate literally. The effect was magical; the woman, who had looked frightened and unhappy, suddenly beamed with smiles, and without any warning she ran toward me, and in an instant I found myself embraced in her loving arms. She pressed me to her bosom, and smothered me with castor-oily kisses, while her greasy ringlets hung upon my face and neck. How long this entertainment would have lasted I cannot tell, but I was obliged to cry "Caffa! Caffa!" (enough! enough!) as it looked improper, and the perfumery was too rich. Fortunately my wife was present, but she did not appear to enjoy it more than I did. My snow-white blouse was soiled and greasy, and for the rest of the day I was a disagreeable compound of smells-- castor oil, tallow, musk, sandal-wood, burnt shells, and Barrake.

Mahomet and Barrake herself, I believe, were the only people who really enjoyed this little event. "Ha!" Mahomet exclaimed, "this is your own fault! You insisted upon speaking kindly, and telling her that she is not a slave; now she thinks that she is one of your WIVES!" This was the real fact; the unfortunate ** Barrake ** had deceived herself. Never having been free, she could not understand the use of freedom unless she was to be a wife. She had understood my little address as a proposal, and of course she was disappointed; but as an action for breach of promise cannot be pressed in the Soudan, poor Barrake, although free, had not the happy rights of a free-born Englishwoman, who can heal her broken heart with a pecuniary plaster, and console herself with damages for the loss of a lover.

We were ready to start, having our party of servants complete, six Tokrooris--Moosa, Abdoolahi, Abderachman, Hassan, Adow, and Hadji Ali, with Mahomet, Wat Gamma, Bacheet, Mahomet secundus (a groom), and Barrake; total, eleven men and the cook.

When half way on our return from Katariff to Wat el Negur, we found the whole country in alarm, Mek Nimmur having suddenly made a foray. He had crossed the Atbara, plundered the district, and driven off large numbers of cattle and camels, after having killed a considerable number of people. No doubt the reports were somewhat exaggerated, but the inhabitants of the district were flying from their villages with their herds, and were flocking to Katariff. We arrived at Wat el Negur on the 3d of December, and we now felt the advantage of our friendship with the good Sheik Achmet, who, being a friend of Mek Nimmur, had saved our effects during our absence. These would otherwise have been plundered, as the robbers had paid him a visit. He had removed our tents and baggage to his own house for protection. Not only had he thus protected our effects, but he had taken the opportunity of delivering the polite message to Mek Nimmur that I had entrusted to his charge--expressing a wish to pay him a visit as a countryman and friend of Mr. Mansfield Parkyns, who had formerly been so well received by his father.

My intention was to examine thoroughly all the great rivers of Abyssinia that were tributaries to the Nile. These were the Settite, Royan, Angrab, Salaam, Rahad, Dinder, and the Blue Nile. If possible, I should traverse the Galla country, and crossing the Blue Nile, I should endeavor to reach the White Nile. But this latter idea I subsequently found impracticable, as it would have interfered with the proper season for my projected journey up the White Nile in search of the sources. The Hamran Arabs were at this time encamped about twenty- five miles from Wat el Negur. I sent a messenger, accompanied by Mahomet, to the sheik, with the firman of the Viceroy, requesting him to supply me with elephant hunters (aggageers).

During the absence of Mahomet I received a very polite message from Mek Nimmur, accompanied by a present of twenty pounds of coffee, with an invitation to pay him a visit. His country lay between the Settite River and the Bahr Salaam; thus without his invitation I might have found it difficult to traverse his territory. So far all went well. I returned my salaams, and sent word that we intended to hunt through the ** Base ** country, after which we should have the honor of passing a few days with him on our road to the river Salaam, at which place we intended to hunt elephants and rhinoceroses.

Mahomet returned, accompanied by a large party of Hamran Arabs, including several hunters, one of whom was Sheik Abou Do Roussoul, the nephew of Sheik Owat. As his name in full was too long, he generally went by the abbreviation "Abou Do." He was a splendid fellow, a little above six feet one, with a light active figure, but exceedingly well-developed muscles. His face was strikingly handsome; his eyes were like those of a giraffe, but the sudden glance of an eagle lighted them up with a flash during the excitement of conversation, which showed little of the giraffe's gentle character. Abou Do was the only tall man of the party; the others were of middle height, with the exception of a little fellow named Jali, who was not above five feet four inches, but wonderfully muscular, and in expression a regular daredevil.

There were two parties of hunters among the Hamran Arabs, one under Abou Do, and the other consisting of four brothers Sherrif. The latter were the most celebrated aggageers among the renowned tribe of the Hamran. Their father and grandfather had been mighty Nimrods, and the broadswords wielded by their strong arms had descended to the men who now upheld the prestige of the ancient blades. The eldest was Taher Sherrif. His second brother, Roder Sherrif, was a very small, active-looking man, with a withered left arm. An elephant had at one time killed his horse, and on the same occasion had driven its sharp tusk through the arm of the rider, completely splitting the limb, and splintering the bone from the elbow-joint to the wrist to such an extent that by degrees the fragments had sloughed away, and the arm had become shrivelled and withered. It now resembled a mass of dried leather twisted into a deformity, without the slightest shape of an arm; this was about fourteen inches in length from the shoulder. The stiff and crippled hand, with contracted fingers, resembled the claw of a vulture.

In spite of his maimed condition, Roder Sherrif was the most celebrated leader in the elephant hunt. His was the dangerous post to ride close to the head of the infuriated animal and provoke the charge, and then to lead the elephant in pursuit, while the aggageers attacked it from behind. It was in the performance of this duty that he had met with the accident, as his horse had fallen over some hidden obstacle and was immediately caught. Being an exceedingly light weight he had continued to occupy this important position in the hunt, and the rigid fingers of the left hand served as a hook, upon which he could hang the reins.

My battery of rifles was now laid upon a mat for examination; they were in beautiful condition, and they excited the admiration of the entire party. The perfection of workmanship did not appear to interest them so much as the size of the bores. They thrust their fingers down each muzzle, until they at last came to the "Baby," when, finding that two fingers could be easily introduced, they at once fell in love with that rifle in particular.

On the 17th of August, accompanied by the German, Florian, we said good-by to our kind friend Sheik Achmet and left Wat el Negur. At Geera, early at daybreak, several Arabs arrived with a report that elephants had been drinking in the river within half an hour's march of our sleeping-place. I immediately started with my men, accompanied by Florian, and we shortly arrived upon the tracks of the herd. I had three Hamran Arabs as trackers, one of whom, Taher Noor, had engaged to accompany us throughout the expedition.

For about eight miles we followed the spoor through high dried grass and thorny bush, until we at length arrived at a dense jungle of kittar--the most formidable of the hooked thorn mimosas. Here the tracks appeared to wander, some elephants having travelled straight ahead, while others had strayed to the right and left. For about two hours we travelled upon the circuitous tracks of the elephants to no purpose, when we suddenly were startled by the shrill trumpeting of one of these animals in the thick thorns, a few hundred yards to our left. The ground was so intensely hard and dry that it was impossible to distinguish the new tracks from the old, which crossed and recrossed in all directions. I therefore decided to walk carefully along the outskirts of the jungle, trusting to find their place of entrance by the fresh broken boughs. In about an hour we had thus examined two or three miles, without discovering a clew to their recent path, when we turned round a clump of bushes, and suddenly came in view of two grand elephants, standing at the edge of the dense thorns. Having our wind, they vanished instantly into the thick jungle. We could not follow them, as their course was down wind; we therefore made a circuit to leeward for about a mile, and finding that the elephants had not crossed in that direction, we felt sure that we must come upon them with the wind in our favor should they still be within the thorny jungle. This was certain, as it was their favorite retreat.

With the greatest labor I led the way, creeping frequently upon my hands and knees to avoid the hooks of the kittar bush, and occasionally listening for a sound. At length, after upward of an hour passed in this slow and fatiguing advance, I distinctly heard the flap of an elephant's ear, shortly followed by the deep guttural sigh of one of those animals, within a few paces; but so dense was the screen of jungle that I could see nothing. We waited for some minutes, but not the slightest sound could be heard; the elephants were aware of danger, and they were, like ourselves, listening attentively for the first intimation of an enemy.

This was a highly exciting moment. Should they charge, there would not be a possibility of escape, as the hooked thorns rendered any sudden movement almost impracticable. In another moment there was a tremendous crash; and with a sound like a whirlwind the herd dashed through the crackling jungle. I rushed forward, as I was uncertain whether they were in advance or retreat. Leaving a small sample of my nose upon a kittar thorn, and tearing my way, with naked arms, through what, in cold blood, would have appeared impassable, I caught sight of two elephants leading across my path, with the herd following in a dense mass behind them. Firing a shot at the leading elephant, simply in the endeavor to check the herd, I repeated with the left-hand barrel at the head of his companion. This staggered him, and threw the main body into confusion. They immediately closed up in a dense mass, and bore everything before them; but the herd exhibited merely an impenetrable array of hind quarters wedged together so firmly that it was impossible to obtain a head or shoulder shot.

I was within fifteen paces of them, and so compactly were they packed that with all their immense strength they could not at once force so extensive a front through the tough and powerful branches of the dense kittar. For about half a minute they were absolutely checked, and they bored forward with all their might in their determination to open a road through the matted thorns. The elastic boughs, bent from their position, sprang back with dangerous force, and would have fractured the skull of any one who came within their sweep. A very large elephant was on the left flank, and for an instant he turned obliquely to the left. I quickly seized the opportunity and fired the "Baby," with an explosive shell, aimed far back in the flank, trusting that it would penetrate beneath the opposite shoulder. The recoil of the "Baby," loaded with ten drams of the strongest powder and a half-pound shell, spun me round like a top. It was difficult to say which was staggered the more severely, the elephant or myself. However, we both recovered, and I seized one of my double rifles, a Reilly No. 10, that was quickly pushed into my hand by my Tokroori, Hadji Ali. This was done just in time, as an elephant from the battled herd turned sharp round, and, with its immense ears cocked, charged down upon us with a scream of rage. "One of us she must have if I miss!"

This was the first downright charge of an African elephant that I had seen, and instinctively I followed my old Ceylon plan of waiting for a close shot. She lowered her head when within about six yards, and I fired low for the centre of the forehead, exactly in the swelling above the root of the trunk. She collapsed to the shot, and fell dead, with a heavy shock, upon the ground. At the same moment the thorny barrier gave way before the pressure of the herd, and the elephants disappeared in the thick jungle, through which it was impossible to follow them.

I had suffered terribly from the hooked thorns, and the men had likewise. This had been a capital trial for my Tokrooris, who had behaved remarkably well, and had gained much confidence by my successful forehead-shot at the elephant when in full charge; but I must confess that this is the only instance in which I have succeeded in killing an African elephant by the front shot, although I have steadily tried the experiment upon subsequent occasions.

We had very little time to examine the elephant, as we were far from home and the sun was already low. I felt convinced that the other elephant could not be far off, after having received the "Baby's" half-pound shell carefully directed, and I resolved to return on the following morning with many people and camels to divide the flesh. It was dark by the time we arrived at the tents, and the news immediately spread through the Arab camp that two elephants had been killed.

On the following morning we started, and upon arrival at the dead elephant we followed the tracks of that wounded by the "Baby." The blood upon the bushes guided us in a few minutes to the spot where the elephant lay dead, at about three hundred yards' distance. The whole day passed in flaying the two animals and cutting off the flesh, which was packed in large gum sacks, with which the camels were loaded. I was curious to examine the effect of the half-pound shell. It had entered the flank on the right side, breaking the rib upon which it had exploded; it had then passed through the stomach and the lower portion of the lungs, both of which were terribly shattered; and breaking one of the fore-ribs on the left side, it had lodged beneath the skin of the shoulder. This was irresistible work, and the elephant had evidently dropped in a few minutes after having received the shell.

A most interesting fact had occurred. I noticed an old wound unhealed and full of matter in the front of the left shoulder. The bowels were shot through, and were green in various places. Florian suggested that it must be an elephant that I had wounded at Wat el Negur; we tracked the course of the bullet most carefully, until we at length discovered my unmistakable bullet of quicksilver and lead, almost uninjured, in the fleshy part of the thigh, imbedded in an unhealed wound. Thus, by a curious chance, upon my first interview with African elephants by daylight, I had killed the identical elephant that I had wounded at Wat el Negur forty-three days before in the dhurra plantation, twenty-eight miles distant!

CHAPTER VII.

The start from Geera--Feats of horsemanship--A curious chase-- Abou Do wins a race--Capturing a young buffalo--Our island camp--Tales of the Base.

We started from Geera on the 23d of December, with our party complete. The Hamran sword-hunters were Abou Do, Jali, and Suleiman. My chief tracker was Taher Noor, who, although a good hunter, was not a professional aggahr, and I was accompanied by the father of Abou Do, who was a renowned "howarti" or harpooner of hippopotami. This magnificent old man might have been Neptune himself. He stood about six feet two, and his grizzled locks hung upon his shoulders in thick, and massive curls, while his deep bronze features could not have been excelled in beauty of outline. A more classical figure I have never beheld than the old Abou Do with his harpoon as he first breasted the torrent, and then landed dripping from the waves to join our party from the Arab camp on the opposite side of the river. In addition to my Tokrooris, I had engaged nine camels, each with a separate driver, of the Hamrans, who were to accompany us throughout the expedition. These people were glad to engage themselves, with their camels included, at one and a half dollars per month, for man and beast as one. We had not sufficient baggage to load five camels, but four carried a large supply of corn for our horses and people.

Hardly were we mounted and fairly started than the monkey-like agility of our aggageers was displayed in a variety of antics, that were far more suited to performances in a circus than to a party of steady and experienced hunters, who wished to reserve the strength of their horses for a trying journey.

Abou Do was mounted on a beautiful Abyssinian horse, a gray; Suleiman rode a rough and inferior-looking beast; while little Jali, who was the pet of the party, rode a gray snare, not exceeding fourteen hands in height, which matched her rider exactly in fire, spirit, and speed. Never was there a more perfect picture of a wild Arab horseman than Jali on his mare. Hardly was he in the saddle than away flew the mare over the loose shingles that formed the dry bed of the river, scattering the rounded pebbles in the air from her flinty Hoofs, while her rider in the vigour of delight threw himself almost under her belly while at full speed, and picked up stones from the ground, which he flung, and again caught as they descended. Never were there more complete Centaurs than these Hamran Arabs ; the horse and man appeared to be one animal, and that of the most elastic nature, that could twist and turn with the suppleness of a snake. The fact of their being separate beings was well proved, however, by the rider's springing to the earth with his drawn sword while the horse was in full gallop over rough and difficult ground, and, clutching the mane, again vaulting into the saddle with the ability of a monkey, without once checking the speed. The fact of being on horseback had suddenly altered the character of these Arabs; from a sedate and proud bearing, they had become the wildest examples of the most savage disciples of Nimrod. Excited by enthusiasm, they shook their naked blades aloft till the steel trembled in their grasp, and away they dashed over rocks, through thorny bush, across ravines, up and down steep inclinations, engaging in a mimic hunt, and going through the various acts supposed to occur in the attack of a furious elephant. I must acknowledge that, in spite of my admiration for their wonderful dexterity, I began to doubt their prudence. I had three excellent horses for my wife and myself; the Hamran hunters had only one for each, and if the commencement were an example of their usual style of horsemanship, I felt sure that a dozen horses would not be sufficient for the work before us. However, it was not the moment to offer advice, as they were simply mad with excitement and delight.

The women raised their loud and shrill yell at parting, and our party of about twenty-five persons, with nine camels, six horses, and two donkeys, exclusive of the German, Florian, with his kicking giraffe-hunter, and attendants, ascended the broken slope that formed the broad valley of the Settite River.

There was very little game in the neighbourhood, as it was completely overrun by the Arabs and their flocks, and we were to march about fifty miles east-south- east before we should arrive in the happy hunting-grounds of the Base country, where we were led to expect great results.

In a day's march through a beautiful country, sometimes upon the high table-land to cut off a bend in the river, at other times upon the margin of the stream in the romantic valley, broken into countless hills and ravines covered with mimosas, we arrived at Ombrega (mother of the thorn), about twenty-four miles from Geera. We soon arranged a resting-place, and cleared away the grass that produced the thorn which had given rise to the name of Ombrega, and in a short time we were comfortably settled for the night. We were within fifty yards of the river, the horses were luxuriating in the green grass that grew upon its banks, and the camels were hobbled, to prevent them from wandering from the protection of the camp-fires, as we were now in the wilderness, where the Base by day and the lion and leopard by night were hostile to man and beast.

We were fast asleep a little after midnight, when we were awakened by the loud barking of the dogs, and by a confusion in the camp. Jumping up on the instant, I heard the dogs, far away in the dark jungles, barking in different directions. One of the goats was gone! A leopard had sprung into the camp, and had torn a goat from its fastening, although tied to a peg, between two men, close to a large fire. The dogs had given chase; but, as usual in such cases, they were so alarmed as to be almost useless. We quickly collected firebrands and searched the jungles, and shortly we arrived where a dog was barking violently. Near this spot we heard the moaning of some animal among the bushes, and upon a search with firebrands we discovered the goat, helpless upon the ground, with its throat lacerated by the leopard. A sudden cry from the dog at a few yards' distance, and the barking ceased.

The goat was carried to the camp where it shortly died. We succeeded in recalling two of the dogs, but the third, which was the best, was missing, having been struck by the leopard. We searched for the body in vain, and concluded that it had been carried off.

The country that we now traversed was so totally uninhabited that it was devoid of all footprints of human beings; even the sand by the river's side, that, like the snow, confessed every print, was free from all traces of man. The Bas-e were evidently absent from our neighbourhood.

We had several times disturbed antelopes during the early portion of the march, and we had just ascended from the rugged slopes of the valley, when we observed a troop of about 100 baboons, which were gathering gum-arabic from the mimosas; upon seeing us, they immediately waddled off. "Would the lady like to have a girrit (baboon)?" exclaimed the ever-excited Jali. Being answered in the affirmative, away dashed the three hunters in full gallop after the astonished apes, who, finding themselves pursued, went off at their best speed. The ground was rough, being full of broken hollows, covered scantily with mimosas, and the stupid baboons, instead of turning to the right into the rugged and steep valley of the Settite, where they would have been secure from the aggageers, kept a straight course before the horses. It was a curious hunt. Some of the very young baboons were riding on their mother's backs; these were now going at their best pace, holding onto their maternal steeds, and looking absurdly humans but in a few minutes, as we closely followed the Arabs, we were all in the midst of the herd, and with great dexterity two of the aggageers, while at full speed, stooped like falcons from their saddles, and seized each a half-grown ape by the back of the neck, and hoisted them upon the necks of the horses. Instead of biting, as I had expected, the astonished captives sat astride of the horses, and clung tenaciously with both arms to the necks of their steeds, screaming with fear.

The hunt was over, and we halted to secured the prisoners. Dismounting, to my surprise the Arabs immediately stripped from a mimosa several thongs of bark, and having tied the baboons by the neck, they gave them a merciless whipping with their powerful coorbatches of hippopotamus hide. It was in vain that I remonstrated against this harsh treatment; they persisted in the punishment. Otherwise they declared that the baboons would bite, but if well-whipped they would become "miskeen"(humble). At length by wife insisted upon mercy, and the unfortunate captives wore an expression of countenance like prisoners about to be led to execution, and they looked imploringly at our faces, in which they evidently discovered some sympathy with their fate. They were quickly placed on horseback before their captors, and once more we continued our journey, highly amused with the little entr' acte.

We had hardly ridden half a mile when I perceived a fine bull tetel standing near a bush a few hundred yards distant. Motioning to the party to halt, I dismounted, and with that the little Fletcher rifle I endeavored to obtain a shot. When within about a hundred and seventy yards, he observed our party, and I was obliged to take the shot, although I could have approached unseen to a closer distance, had his attention not been attracted by the noise of the horses. He threw his head up preparatory to starting off, and he was just upon the move as I touched the trigger. He fell like a stone to the shot, but almost immediately he regained his feet and bounded off, receiving a bullet from the second barrel without a flinch. In full speed he rushed away across the party of aggageers about three hundred yards distant.

Out dashed Abou Do from the ranks on his active gray horse, and away he flew after the wounded tetel, his long hair floating in the wind, his naked sword in hand, and his heels digging into the flanks of his horse, as though armed with spurs in the last finish of a race. It was a beautiful course. Abou Do hunted like a cunning greyhound; the tetel turned, and, taking advantage of the double, he cut off the angle; succeeding by the manoeuvre, he again followed at tremendous speed over the numerous inequalities of the ground, gaining in the race until he was within twenty yards of the tetel, when we lost sight of both game and hunter in the thick bushes. By this time I had regained my horse, that was brought to meet me, and I followed to the spot, toward which my wife and the aggageers, encumbered with the unwilling apes, were already hastening. Upon arrival I found, in high yellow grass beneath a large tree, the tetel dead, and Abou Do wiping his bloody sword, surrounded by the foremost of the party. He had hamstrung the animal so delicately that the keen edge of the blade was not injured against the bone. My two bullets had passed through the tetel. The first was too high, having entered above the shoulder--this had dropped the animal for a moment; the second was through the flank.

The Arabs now tied the baboons to trees, and employed themselves in carefully skinning the tetel so as to form a sack from the hide. They had about half finished the operation, when we were disturbed by a peculiar sound at a considerable distance in the jungle, which, being repeated, we knew to be the cry of buffaloes. In an instant the tetel was neglected, the aggageers mounted their horses, and leaving my wife with a few men to take charge of the game, accompanied by Florian we went in search of the buffaloes. This part of the country was covered with grass about nine feet high, that was reduced to such extreme dryness that the stems broke into several pieces like glass as we brushed through it. The jungle was open, composed of thorny mimosas at such wide intervals that a horse could be ridden at considerable speed if accustomed to the country. Altogether it was the perfection of ground for shooting, and the chances were in favour of the rifle.

We had proceeded carefully about half a mile when I heard a rustling in the grass, and I shortly perceived a bull buffalo standing alone beneath a tree, close to the sandy bed of a dried stream, which was about a hundred yards distant, between us and the animal. The grass had been entirely destroyed by the trampling of a large herd. I took aim at the shoulder with one of my No. 10 Reilly rifles, and the buffalo rushed forward at the shot, and fell about a hundred paces beyond in the bush. At the report of the shot, the herd, that we had not observed, which had been lying upon the sandy bed of the stream, rushed past us with a sound like thunder, in a cloud of dust raised by several hundreds of large animals in full gallop. I could hardly see them distinctly, and I waited for a good chance, when presently a mighty bull separated from the rest, and gave me a fair shoulder-shot. I fired a little too forward, and missed the shoulder; but I made a still better shot by mistake, as the Reilly bullet broke the spine through the neck, and dropped him dead. Florian, poor fellow, had not the necessary tools for the work, and one of his light guns produced no effect.

Now came the time for the aggageers. Away dashed Jali op his fiery mare, closely followed by Abou Do and Suleiman, who in a few instants were obscured in the cloud of dust raised by the retreating buffaloes. As soon as I could mount my horse that had been led behind me, I followed at full speed, and, spurring hard, I shortly came in sight of the three aggageers, not only in the dust, but actually among the rear buffaloes of the herd. Suddenly, Jali almost disappeared from the saddle as he leaned forward with a jerk and seized a fine young buffalo by the tail. In a moment Abou Do and Suleiman sprang from their horses, and I arrived just in time to assist them in securing a fine little bull about twelve hands high, whose horns were six or seven inches long. A pretty fight we had with the young Hercules. The Arabs stuck to him like bulldogs, in spite of his tremendous struggles, and Florian, with other men, shortly arriving, we secured him by lashing his legs together with our belts until impromptu ropes could be made with mimosa bark.

I now returned to the spot where we had left my wife and the tetel. I found her standing about fifty yards from the spot with a double rifle cocked, awaiting an expected charge from one of the buffaloes that, separated from the herd, had happened to rush in her direction.

Mahomet had been in an awful fright, and was now standing secure behind his mistress. I rode through the grass with the hope of getting a shot, but the animal had disappeared. We returned to the dead tetel and to our captive baboons; but times had changed since we had left them. One had taken advantage of our absence, and, having bitten through his tether, had escaped. The other had used force instead of cunning, and, in attempting to tear away from confinement, had strangled himself with the slip-knot of the rope.

We now pushed ahead, and at 5 P.M. we arrived at the spot on the margin of the Settite River at which we were to encamp for some time. For many miles on either side the river was fringed with dense groves of the green nabbuk, but upon the east bank an island had been formed of about three hundred acres. This was a perfect oasis of verdure, covered with large nabbuk trees, about thirty feet high, and forming a mixture of the densest coverts, with small open glades of rich but low herbage. To reach this island, upon which we were to encamp, it was necessary to cross the arm of the river, that was now dry, with the exception of deep pools, in one of which we perceived a large bull buffalo drinking, just as we descended the hill. As this would be close to the larder, I stalked to within ninety yards, and fired a Reilly No. 10 into his back, as his head inclined to the water. For the moment he fell upon his knees, but recovering immediately, he rushed up the steep bank of the island, receiving the ball from my left-hand barrel between his shoulders, and disappeared in the dense covert of green nabbuk on the margin. As we were to camp within a few yards of the spot, he was close to home; therefore, having crossed the river, we carefully followed the blood tracks through the jungle. But, after having pushed our way for about twenty paces through the dense covert, I came to the wise conclusion that it was not the place for following a wounded buffalo, and that we should find him dead on the next morning.

A few yards upon our right hand was a beautiful open glade, commanding a view of the river, and surrounded by the largest nabbuk trees, that afforded a delightful shade in the midst of the thick covert. This was a spot that in former years had been used by the aggageers as a camp, and we accordingly dismounted and turned the horses to graze upon the welcome grass. Each horse was secured to a peg by a long leathern thong, as the lions in this neighbourhood were extremely dangerous, having the advantage of thick and opaque jungle.

We employed ourselves until the camels should arrive in cutting thorn branches and constructing a zareeba or fenced camp, to protect our animals during the night from the attack of wild beasts. I also hollowed out a thick green bush to form an arbour, as a retreat during the heat of the day, and in a short space of time we were prepared for the reception of the camels and effects. The river had cast up immense stores of dry wood; this we had collected, and by the time the camels arrived with the remainder of our party after dark, huge fires were blazing high in air, the light of which had guided them direct to our camp. They were heavily laden with meat, which is the Arab's great source of happiness; therefore in a few minutes the whole party was busily employed in cutting the flesh into long thin strips to dry. These were hung in festoons over the surrounding trees, while the fires were heaped with tidbits of all descriptions. I had chosen a remarkably snug position for ourselves; the two angareps (stretchers) were neatly arranged in the middle of a small open space free from overhanging boughs; near these blazed a large fire, upon which were roasting a row of marrow-bones of buffalo and tetel, while the table was spread with a clean cloth and arranged for dinner.

The woman Barrak, who had discovered with regret that she was not a wife but a servant, had got over the disappointment, and was now making dhurra cakes upon the doka. This is a round earthenware tray about eighteen inches in diameter, which, supported upon three stones or lumps of earth, over a fire of glowing embers, forms a hearth. Slices of liver, well peppered with cayenne and salt, were grilling on the gridiron, and we were preparing to dine, when a terrific roar within a hundred and fifty yards informed us that a lion was also thinking of dinner. A confusion of tremendous roars proceeding from several lions followed the first round, and my aggageers quietly remarked, "There is no danger for the horses tonight; the lions have found your wounded buffalo!"

Such a magnificent chorus of bass voices I had never heard. The jungle cracked, as with repeated roars they dragged the carcass of the buffalo through the thorns to the spot where they intended to devour it. That which was music to our ears was discord to those of Mahomet, who with terror in his face came to us and exclaimed, "Master, what's that? What for master and the missus come to this bad country? That's one bad kind will eat the missus in the night! Perhaps he come and eat Mahomet!" This afterthought was too much for him, and Bacheet immediately comforted him by telling the most horrible tales of death and destruction that had been wrought by lions, until the nerves of Mahomet were completely unhinged.

This was a signal for story-telling, when suddenly the aggageers changed the conversation by a few tales of the Bas-e natives, which so thoroughly eclipsed the dangers of wild beasts that in a short time the entire party would almost have welcomed a lion, provided he would have agreed to protect them from the Bas-e. In this very spot where we were then camped, a party of Arab hunters had, two years previous, been surprised at night and killed by the Bas-e, who still boasted of the swords that they possessed as spoils from that occasion. The Bas-e knew this spot as the favorite resting-place of the Hamran hunting-parties, and they might be not far distant NOW, as we were in the heart of their country. This intelligence was a regular damper to the spirits of some of the party. Mahomet quietly retired and sat down by Barrak, the ex-slave woman, having expressed a resolution to keep awake every hour that he should be compelled to remain in that horrible country. The lions roared louder and louder, but no one appeared to notice such small thunder; all thoughts were fixed upon the Bas-e, so thoroughly had the aggageers succeeded in frightening not only Mahomet, but also our Tokrooris.

CHAPTER VIII.

The elephant trumpets--Fighting an elephant with swords--The forehead-shot--Elephants in a panic--A superb old Neptune-- The harpoon reaches its aim--Death of the hippopotamus-- Tramped by an elephant.

The aggageers started before daybreak in search of elephants. They soon returned, and reported the fresh tracks of a herd, and begged me to lose no time in accompanying them, as the elephants might retreat to a great distance. There was no need for this advice. In a few minutes my horse Tetel was saddled, and my six Tokrooris and Bacheet, with spare rifles, were in attendance. Bacheet, who had so ingloriously failed in his first essay at Wat el Negur, had been so laughed at by the girls of the village for his want of pluck that he had declared himself ready to face the devil rather than the ridicule of the fair sex; and, to do him justice, he subsequently became a first-rate lad in moments of danger.

The aggageers were quickly mounted. It was a sight most grateful to a sportsman to witness the start of these superb hunters, who with the sabres slung from the saddle-bow, as though upon an every-day occasion, now left the camp with these simple weapons, to meet the mightiest animal of creation in hand-to-hand conflict. The horses' hoofs clattered as we descended the shingly beach, and forded the river shoulder-deep, through the rapid current, while those on foot clung to the manes of the horses and to the stirrup-leathers to steady themselves over the loose stones beneath.

Tracking was very difficult. As there was a total absence of rain, it was next to impossible to distinguish the tracks of two days' date from those most recent upon the hard and parched soil. The only positive clew was the fresh dung of the elephants, and this being deposited at long intervals rendered the search extremely tedious. The greater part of the day passed in useless toil, and, after fording the river backward and forward several times, we at length arrived at a large area of sand in the bend of the stream, that was evidently overflowed when the river was full. This surface of many acres was backed by a forest of large trees. Upon arrival at this spot the aggageers, who appeared to know every inch of the country, declared that, unless the elephants had gone far away, they must be close at hand, within the forest. We were speculating upon the direction of the wind, when we were surprised by the sudden trumpeting of an elephant, that proceeded from the forest already declared to be the covert of the herd. In a few minutes later a fine bull elephant marched majestically from the jungle upon the large area of sand, and proudly stalked direct toward the river.

At that time we were stationed under cover of a high bank of sand that had been left by the retiring river in sweeping round an angle. We immediately dismounted, and remained well concealed. The question of attack was quickly settled. The elephant was quietly stalking toward the water, which was about three hundred paces distant from the jungle. This intervening space was heavy dry sand, that had been thrown up by the stream in the sudden bend of the river, which, turning from this point at a right angle, swept beneath a perpendicular cliff of conglomerate rock formed of rounded pebbles cemented together.

I proposed that we should endeavor to stalk the elephant, by creeping along the edge of the river, under cover of a sand-bank about three feet high, and that, should the rifles fail, the aggageers should come on at full gallop and cut off his retreat from the jungle; we should then have a chance for the swords.

Accordingly I led the way, followed by Hadji Ali, my head Tokroori, with a rifle, while I carried the "Baby." Florian accompanied us. Having the wind fair, we advanced quickly for about half the distance, at which time we were within a hundred and fifty yards of the elephant, who had just arrived at the water and had commenced drinking. We now crept cautiously toward him. The sand-bank had decreased to a height of about two feet, and afforded very little shelter. Not a tree or bush grew upon the surface of the barren sand, which was so deep that we sank nearly to the ankles at every footstep. Still we crept forward, as the elephant alternately drank and then spouted the water in a shower over his colossal form; but just as we arrived within about fifty yards he happened to turn his head in our direction, and immediately perceived us. He cocked his enormous ears, gave a short trumpeting, and for an instant wavered in his determination whether to attack or fly; but as I rushed toward him with a shout, he turned toward the jungle, and I immediately fired a steady shot at the shoulder with the "Baby." As usual, the fearful recoil of the rifle, with a half-pound shell and twelve drams of powder, nearly threw me backward; but I saw the mark upon the elephant's shoulder, in an excellent line, although rather high. The only effect of the shot was to send him off at great speed toward the jungle. At the same moment the three aggageers came galloping across the sand like greyhounds in a course, and, judiciously keeping parallel with the jungle, they cut off his retreat, and, turning toward the elephant, confronted him, sword in hand.

At once the furious beast charged straight at the enemy. But now came the very gallant but foolish part of the hunt. Instead of leading the elephant by the flight of one man and horse, according to their usual method, all the aggageers at the same moment sprang from their saddles, and upon foot in the heavy sand they attacked the elephant with their swords.

In the way of sport I never saw anything so magnificent or so absurdly dangerous. No gladiatorial exhibition in the Roman arena could have surpassed this fight. The elephant was mad with rage, and nevertheless he seemed to know that the object of the hunters was to get behind him. This he avoided with great dexterity, turning as it were upon a pivot with extreme quickness, and charging headlong, first at one and then at another of his assailants, while he blew clouds of sand in the air with his trunk, and screamed with fury. Nimble as monkeys, nevertheless the aggageers could not get behind him. In the folly of excitement they had forsaken their horses, which had escaped from the spot. The depth of the loose sand was in favor of the elephant, and was so much against the men that they avoided his charges with extreme difficulty. It was only by the determined pluck of all three that they alternately saved each other, as two invariably dashed in at the flanks when the elephant charged the third, upon which the wary animal immediately relinquished the chase and turned round upon his pursuers. During this time I had been laboring through the heavy sand, and shortly after I arrived at the fight the elephant charged directly through the aggageers, receiving a shoulder-shot from one of my Reilly No. 10 rifles, and at the same time a slash from the sword of Abou Do, who with great dexterity and speed had closed in behind him, just in time to reach the leg. Unfortunately, he could not deliver the cut in the right place, as the elephant, with increased speed, completely distanced the aggageers, then charged across the deep sand and reached the jungle. We were shortly upon his tracks, and after running about a quarter of a mile he fell dead in a dry watercourse. His tusks were, like those of most Abyssinian elephants, exceedingly short, but of good thickness.

Some of our men, who had followed the runaway horses, shortly returned and reported that during our fight with the bull they had heard other elephants trumpeting in the dense nabbuk jungle near the river. We all dismounted, and sent the horses to a considerable distance, lest they should by some noise disturb the elephants. We shortly heard a crackling in the jungle on our right, and Jali assured us that, as he had expected, the elephants were slowly advancing along the jungle on the bank of the river, and would pass exactly before us. We waited patiently in the bed of the river, and the crackling in the jungle sounded closer as the herd evidently approached. The strip of thick thorny covert that fringed the margin was in no place wider than half a mile; beyond that the country was open and park-like, but at this season it was covered with parched grass from eight to ten feet high. The elephants would, therefore, most probably remain in the jungle until driven out.

In about a quarter of an hour we knew by the noise in the jungle, about a hundred yards from the river, that the elephants were directly opposite to us. I accordingly instructed Jali to creep quietly by himself into the bush and to bring me information of their position. To this he at once agreed.

In three or four minutes he returned. He declared it impossible to use the sword, as the jungle was so dense that it would check the blow; but that I could use the rifle, as the elephants were close to us--he had seen three standing together, between us and the main body of the herd. I told Jali to lead me directly to the spot, and, followed by Florian and the aggageers, with my gun-bearers, I kept within a foot of my dependable little guide, who crept gently into the jungle. This was exceedingly thick, and quite impenetrable, except in the places where elephants and other heavy animals had trodden numerous alleys. Along one of these narrow passages we stealthily advanced, until Jali stepped quietly on one side and pointed with his finger. I immediately observed two elephants looming through the thick bushes about eight paces from me. One offered a temple-shot, which I quickly took with a Reilly No. 10, and floored it on the spot. The smoke hung so thickly that I could not see distinctly enough to fire my second barrel before the remaining elephant had turned; but Florian, with a three-ounce steel-tipped bullet, by a curious shot at the hind-quarters, injured the hip joint to such an extent that we could more than equal the elephant in speed.

In a few moments we found ourselves in a small open glade in the middle of the jungle, close to the stern of the elephant we were following. I had taken a fresh rifle, with both barrels loaded, and hardly had I made the exchange when the elephant turned suddenly and charged. Determined to try fairly the forehead-shot, I kept my ground, and fired a Reilly No. 10, quicksilver and lead bullet, exactly in the centre, when certainly within four yards. The only effect was to make her stagger backward, when, in another moment, with her immense ears thrown forward, she again rushed on. This was touch-and-go; but I fired my remaining barrel a little lower than the first shot. Checked in her rush, she backed toward the dense jungle, throwing her trunk about and trumpeting with rage. Snatching the Ceylon No. 10 from one of my trusty Tokrooris (Hassan), I ran straight at her, took a most deliberate aim at the forehead, and once more fired. The only effect was a decisive charge; but before I fired my last barrel Jali rushed in, and, with one blow of his sharp sword, severed the back sinew. She was utterly helpless in the same instant. Bravo, Jali! I had fired three beautifully correct shots with No. 10 bullets and seven drams of powder in each charge. These were so nearly together that they occupied a space in her forehead of about three inches, and all had failed to kill! There could no longer be any doubt that the forehead-shot at an African elephant could not be relied upon, although so fatal to the Indian species. This increased the danger tenfold, as in Ceylon I had generally made certain of an elephant by steadily waiting until it was close upon me.

I now reloaded my rifles, and the aggageers quitted the jungle to remount their horses, as they expected the herd had broken cover on the other side of the jungle, in which case they intended to give chase, and, if possible, to turn them back into the covert and drive them toward the guns. We accordingly took our stand in the small open glade, and I lent Florian one of my double rifles, as he was only provided with one single-barrelled elephant gun. I did not wish to destroy the prestige of the rifles by hinting to the aggageers that it would be rather awkward for us to receive the charge of the infuriated herd, as the foreheads were invulnerable; but inwardly I rather hoped that they would not come so directly upon our position as the aggageers wished.

About a quarter of an hour passed in suspense, when we suddenly heard a chorus of wild cries of excitement on the other side of the jungle, raised by the aggageers, who had headed the herd and were driving them back toward us. In a few minutes a tremendous crashing in the jungle, accompanied by the occasional shrill scream of a savage elephant and the continued shouts of the mounted aggageers, assured us that they were bearing down exactly upon our direction. They were apparently followed even through the dense jungle by the wild and reckless Arabs. I called my men close together, told them to stand fast and hand me the guns quickly, and we eagerly awaited the onset that rushed toward us like a storm.

On they came, tearing everything before them. For a moment the jungle quivered and crashed; a second later, and, headed by an immense elephant, the herd thundered down upon us. The great leader came directly at me, and was received with right and left in the forehead from a Reilly No. 10 as fast as I could pull the triggers. The shock made it reel backward for an instant, and fortunately turned it and the herd likewise. My second rifle was beautifully handed, and I made a quick right and left at the temples of two fine elephants, dropping them both stone dead. At this moment the "Baby" was pushed into my hand by Hadji Ali just in time to take the shoulder of the last of the herd, who had already charged headlong after his comrades and was disappearing in the jungle. Bang! went the "Baby;" round I spun like a weathercock, with the blood pouring from my nose, as the recoil had driven the sharp top of the hammer deep into the bridge. My "Baby" not only screamed, but kicked viciously. However, I knew that the elephant must be bagged, as the half-pound shell had been aimed directly behind the shoulder.

In a few minutes the aggageers arrived. They were bleeding from countless scratches, as, although naked with the exception of short drawers, they had forced their way on horseback through the thorny path cleft by the herd in rushing through the jungle. Abou Do had blood upon his sword. They had found the elephants commencing a retreat to the interior of the country, and they had arrived just in time to turn them. Following them at full speed, Abou Do had succeeded in overtaking and slashing the sinew of an elephant just as it was entering the jungle. Thus the aggageers had secured one, in addition to Florian's elephant that had been slashed by Jali. We now hunted for the "Baby's" elephant, which was almost immediately discovered lying dead within a hundred and fifty yards of the place where it had received the shot. The shell had entered close to the shoulder, and it was extraordinary that an animal should have been able to travel so great a distance with a wound through the lungs by a shell that had exploded within the body.

We had done pretty well. I had been fortunate in bagging four from this herd, in addition to the single bull in the morning; total, five. Florian had killed one and the aggageers one; total, seven elephants. One had escaped that I had wounded in the shoulder, and two that had been wounded by Florian. The aggageers were delighted, and they determined to search for the wounded elephants on the following day, as the evening was advancing, and we were about five miles from camp.

At daybreak the next morning the aggageers in high glee mounted their horses, and with a long retinue of camels and men, provided with axes and knives, together with large gum sacks to contain the flesh, they quitted the camp to cut up the numerous elephants. As I had no taste for this disgusting work, I took two of my Tokrooris, Hadji Ali and Hassan, and, accompanied by old Abou Do, the father of the sheik, with his harpoon, we started along the margin of the river in quest of hippopotami.

The harpoon for hippopotamus and crocodile hunting is a piece of soft steel about eleven inches long, with a narrow blade or point of about three quarters of an inch in width and a single but powerful barb. To this short and apparently insignificant weapon a strong rope is secured, about twenty feet in length, at the extremity of which is a buoy or float, as large as a child's head, formed of an extremely light wood called ambatch (Aanemone mirabilis) that is of about half the specific gravity of cork. The extreme end of the short harpoon is fixed in the point of a bamboo about ten feet long, around which the rope is twisted, while the buoy end is carried in the left hand.

The old Abou Do, being resolved upon work, had divested himself of his tope or toga before starting, according to the general custom of the aggageers, who usually wear a simple piece of leather wound round the loins when hunting; but, I believe in respect for our party, they had provided themselves with a garment resembling bathing drawers, such as are worn in France, Germany, and other civilized countries. But the old Abou Do had resisted any such innovation, and he accordingly appeared with nothing on but his harpoon; and a more superb old Neptune I never beheld. He carried this weapon in his hand, as the trident with which the old sea-god ruled the monsters of the deep; and as the tall Arab patriarch of threescore years and ten, with his long gray locks flowing over his brawny shoulders, stepped as lightly as a goat from rock to rock along the rough margin of the river, I followed him in admiration.

After walking about two miles we noticed a herd of hippopotami in a pool below a rapid. This was surrounded by rocks, except upon one side, where the rush of water had thrown up a bank of pebbles and sand. Our old Neptune did not condescend to bestow the slightest attention when I pointed out these animals; they were too wide awake; but he immediately quitted the river's bed, and we followed him quietly behind the fringe of bushes upon the border, from which we carefully examined the water.

About half a mile below this spot, as we clambered over the intervening rocks through a gorge which formed a powerful rapid, I observed, in a small pool just below the rapid, the immense head of a hippopotamus close to a perpendicular rock that formed a wall to the river, about six feet above the surface. I pointed out the hippo to old Abou Do, who had not seen it. At once the gravity of the old Arab disappeared, and the energy of the hunter was exhibited as he motioned us to remain, while he ran nimbly behind the thick screen of bushes for about a hundred and fifty yards below the spot where the hippo was unconsciously basking, with his ugly head above the surface. Plunging into the rapid torrent, the veteran hunter was carried some distance down the stream; but, breasting the powerful current, he landed upon the rocks on the opposite side, and, retiring to some distance from the river, he quickly advanced toward the spot beneath which the hippopotamus was lying. I had a fine view of the scene, as I was lying concealed exactly opposite the hippo, who had disappeared beneath the water.

Abou Do now stealthily approached the ledge of rock beneath which he had expected to see the head of the animal. His long, sinewy arm was raised, with the harpoon ready to strike, as he carefully advanced. At length he reached the edge of the perpendicular rock. The hippo had vanished, but, far from exhibiting surprise, the old Arab remained standing on the sharp edge, unchanged in attitude. No figure of bronze could have been more rigid than that of the old river-king as he stood erect upon the rock with the left foot advanced and the harpoon poised in his ready right hand above his head, while in the left he held the loose coils of rope attached to the ambatch buoy. For about three minutes he stood like a statue, gazing intently into the clear and deep water beneath his feet. I watched eagerly for the reappearance of the hippo; the surface of the water was still barren, when suddenly the right arm of the statue descended like lightning, and the harpoon shot perpendicularly into the pool with the speed of an arrow. What river-fiend answered to the summons? In an instant an enormous pair of open jaws appeared, followed by the ungainly head and form of the furious hippopotamus, who, springing half out of the water, lashed the river into foam, and, disdaining the concealment of the deep pool, charged straight up the violent rapids. With extraordinary power he breasted the descending stream, gaining a footing in the rapids, about five feet deep. He ploughed his way against the broken waves, sending them in showers of spray upon all sides, and, upon gaining broader shallows, tore along through the water, with the buoyant float hopping behind him along the surface, until he landed from the river, started at full gallop along the dry shingly bed, and at length disappeared in the thorny nabbuk jungle.

I never could have imagined that so unwieldy an animal could have exhibited such speed; no man would have had a chance of escape, and it was fortunate for our old Neptune that he was secure upon the high ledge of rock; for if he had been in the path of the infuriated beast there would have been an end of Abou Do. The old man plunged into the deep pool just quitted by the hippo and landed upon our side, while in the enthusiasm of the moment I waved my cap above my head and gave him a British cheer as he reached the shore. His usually stern features relaxed into a grim smile of delight: this was one of those moments when the gratified pride of the hunter rewards him for any risks. I congratulated him upon his dexterity; but much remained to be done. I proposed to cross the river, and to follow upon the tracks of the hippopotamus, as I imagined that the buoy and rope would catch in the thick jungle, and that we should find him entangled in the bush; but the old hunter gently laid his hand upon my arm and pointed up the bed of the river, explaining that the hippo would certainly return to the water after a short interval.

In a few minutes later, at a distance of nearly half a mile, we observed the hippo emerge from the jungle and descend at full trot to the bed of the river, making direct for the first rocky pool in which we had noticed the herd of hippopotami. Accompanied by the old howarti (hippo hunter), we walked quickly toward the spot. He explained to me that I must shoot the harpooned hippo, as we should not be able to secure him in the usual method by ropes, as nearly all our men were absent from camp, disposing of the dead elephants.

Upon reaching the pool, which was about a hundred and thirty yards in diameter, we were immediately greeted by the hippo, who snorted and roared as we approached, but quickly dived, and the buoyant float ran along the surface, directing his course in the same manner as the cork of a trimmer marks that of a pike upon the hook. Several times he appeared, but as he invariably faced us I could not obtain a favorable shot; I therefore sent the old hunter round the pool, and he, swimming the river, advanced to the opposite side and attracted the attention of the hippo, who immediately turned toward him. This afforded me a good chance, and I fired a steady shot behind the ear, at about seventy yards, with a single-barrelled rifle. As usual with hippopotami, whether dead or alive, he disappeared beneath the water at the shot. The crack of the ball and the absence of any splash from the bullet told me that he was hit; the ambatch float remained perfectly stationary upon the surface. I watched it for some minutes--it never moved. Several heads of hippopotami appeared and vanished in different directions, but the float was still; it marked the spot where the grand old bull lay dead beneath.

I shot another hippo, that I thought must be likewise dead; and, taking the time by my watch, I retired to the shade of a tree with Hassan, while Hadji Ali and the old hunter returned to camp for assistance in men and knives, etc.

In a little more than an hour and a half, two objects like the backs of turtles appeared above the surface. These were the flanks of the two hippos. A short time afterward the men arrived, and, regardless of crocodiles, they swam toward the bodies. One was towed directly to the shore by the rope attached to the harpoon, the other was secured by a long line and dragged to the bank of clean pebbles. We had now a good supply of food, which delighted our people.

I returned to the camp, and several hours elapsed, but none of the aggageers returned, and neither had we received any tidings of our people and camels that had left us at daybreak to search for the dead elephants. Fearing that some mishap might have occurred in a collision with the Bas-e, I anxiously looked out for some sign of the party. At about 4 P.M. I observed far up the bed of the river several men, some mounted and others upon foot, while one led a camel with a curious-looking load. Upon a nearer approach I could distinguish upon the camel's back some large object that was steadied by two men, one of whom walked on either side. I had a foreboding that something was wrong, and in a few minutes I clearly perceived a man lying upon a make-shift litter, carried by the camel, while the Sheik Abou Do and Suleiman accompanied the party upon horseback; a third led Jali's little gray mare.

They soon arrived beneath the high bank of the river upon which I stood. Poor little Jali, my plucky and active ally, lay, as I thought, dead upon the litter. We laid him gently upon my angarep, which I had raised by four men, so that we could lower him gradually from the kneeling camel, and we carried him to the camp, about thirty yards distant. He was faint, and I poured some essence of peppermint (the only spirits I possessed) down his throat, which quickly revived him. His thigh was broken about eight inches above the knee, but fortunately it was a simple fracture.

Abou Do now explained the cause of the accident. While the party of camel, men and others were engaged in cutting up the dead elephants, the three aggageers had found the track of a bull that had escaped wounded. In that country, where there was no drop of water upon the east bank of the Settite for a distance of sixty or seventy miles to the river Gash, an elephant, if wounded, was afraid to trust itself to the interior. One of our escaped elephants had therefore returned to the thick jungle, and was tracked by the aggageers to a position within two or three hundred yards of the dead elephants. As there were no guns, two of the aggageers, utterly reckless of consequences, resolved to ride through the narrow passages formed by the large game, and to take their chance with the elephant, sword in hand. Jali, as usual, was the first to lead, and upon his little gray mare he advanced with the greatest difficulty through the entangled thorns, broken by the passage of heavy game; to the right and left of the passage it was impossible to move. Abou Do had wisely dismounted, but Suleiman followed Jali. Upon arriving within a few yards of the elephant, which was invisible in the thick thorns, Abou Do crept forward on foot, and discovered it standing with ears cocked, evidently waiting for the attack. As Jali followed on his light gray mare, the elephant immediately perceived the white color and at once charged forward. Escape was next to impossible. Jali turned his snare sharply around, and she bounded off; but, caught in the thorns, the mare fell, throwing her rider in the path of the elephant that was within a few feet behind, in full chase. The mare recovered herself in an instant, and rushed away; the elephant, attracted by the white color of the animal, neglected the man, upon whom it trod in the pursuit, thus breaking his thigh. Abou Do, who had been between the elephant and Jali, had wisely jumped into the thick thorns, and, as the elephant passed him, he again sprang out behind and followed with his drawn sword, but too late to save Jali, as it was the affair of an instant.

Jumping over Jali's body, he was just in time to deliver a tremendous cut at the hind leg of the elephant, that must otherwise have killed both horses and probably Suleiman also, as the three were caught in a cul de sac, in a passage that had no outlet, and were at the elephant's mercy.

Abou Do seldom failed. It was a difficult feat to strike correctly in the narrow jungle passage with the elephant in full speed; but the blow was fairly given, and the back sinew was divided. Not content with the success of the cut, he immediately repeated the stroke upon the other leg, as he feared that the elephant, although disabled from rapid motion, might turn and trample Jali. The extraordinary dexterity and courage required to effect this can hardly be appreciated by those who have never hunted a wild elephant; but the extreme agility, pluck, and audacity of these Hamran sword-hunters surpass all feats that I have ever witnessed.

I set Jali's broken thigh and attended to him for four days. He was a very grateful but unruly patient, as he had never been accustomed to remain quiet. At the end of that time we arranged an angarep comfortably upon a camel, upon which he was transported to Geera, in company with a long string of camels, heavily laden with dried meat and squares of hide for shields, with large bundles of hippopotamus skin for whip-making, together with the various spoils of the chase. Last but not least were numerous leathern pots of fat that had been boiled down from elephants and hippopotami.

The camels were to return as soon as possible with supplies of corn for our people and horses. Another elephant-hunter was to be sent to us in the place of Jali, but I felt that we had lost our best man.

CHAPTER IX.

Fright of the Tokrooris--Deserters who didn't desert--Arrival of the Sherrif brothers--Now for a tally-ho!--On the heels of the rhinoceroses--The Abyssinian rhinoceros--Every man for himself.

Although my people had been in the highest spirits up to this time, a gloom had been thrown over the party by two causes--Jali's accident and the fresh footmarks of the Bas-e that had been discovered upon the sand by the margin of the river. The aggageers feared nothing, and if the Bas-e had been legions of demons they would have faced them, sword in hand, with the greatest pleasure. But my Tokrooris, who were brave in some respects, had been so cowed by the horrible stories recounted of these common enemies at the nightly camp-fires by the Hamran Arabs, that they were seized with panic and resolved to desert en masse and return to Katariff, where I had originally engaged them, and at which place they had left their families.

In this instance the desertion of my Tokrooris would have been a great blow to my expedition, as it was necessary to have a division of parties. I had the Tokrooris, Jaleens, and Hamran Arabs. Thus they would never unite together, and I was certain to have some upon my side in a difficulty. Should I lose the Tokrooris, the Hamran Arabs would have the entire preponderance.

The whole of my Tokrooris formed in line before me and my wife, just as the camels were about to leave. Each man had his little bundle prepared for starting on a journey. Old Moosa was the spokesman. He said that they were all very sorry; that they regretted exceedingly the necessity of leaving us, but some of them were sick, and they would only be a burden to the expedition; that one of them was bound upon a pilgrimage to Mecca, and that God would punish him should he neglect this great duty; others had not left any money with their families in Katariff, that would starve in their absence. (I had given them an advance of wages, when they engaged at Katariff, to provide against this difficulty.) I replied: "My good fellows, I am very sorry to hear all this, especially as it comes upon me so suddenly; those who are sick stand upon one side" (several invalids, who looked remarkably healthy, stepped to the left). "Who wishes to go to Mecca?" Abderachman stepped forward (a huge specimen of a Tokroori, who went by the nickname of "El Jamoos" or the buffalo). "Who wishes to remit money to his family, as I will send it and deduct it from his wages?" No one came forward. During the pause I called for pen and paper, which Mahomet brought. I immediately commenced writing, and placed the note within an envelope, which I addressed and gave to one of the camel-drivers. I then called for my medicine-chest, and having weighed several three-grain doses of tartar emetic, I called the invalids, and insisted upon their taking the medicine before they started, or they might become seriously ill upon the road, which for three days' march was uninhabited. Mixed with a little water the doses were swallowed, and I knew that the invalids were safe for that day, and that the others would not start without them.

I now again addressed my would-be deserters: "Now, my good fellows, there shall be no misunderstanding between us, and I will explain to you how the case stands. You engaged yourselves to me for the whole journey, and you received an advance of wages to provide for your families during your absence. You have lately filled yourselves with meat, and you have become lazy; you have been frightened by the footprints of the Bas-e; thus you wish to leave the country. To save yourselves from imaginary danger, you would forsake my wife and myself, and leave us to a fate which you yourselves would avoid. This is your gratitude for kindness; this is the return for my confidence, when without hesitation I advanced you money. Go! Return to Katariff to your families! I know that all the excuses you have made are false. Those who declare themselves to be sick, Inshallah (please God), shall be sick. You will all be welcomed upon your arrival at Katariff. In the letter I have written to the Governor, inclosing your names, I have requested him to give each man upon his appearance FIVE HUNDRED LASHES WITH THE COORBATCH, FOR DESERTION, and to imprison him until my return."

Checkmate! My poor Tokrooris were in a corner, and in their great dilemma they could not answer a word. Taking advantage of this moment of confusion, I called forward "the buffalo," Abderachman, as I had heard that he really had contemplated a pilgrimage to Mecca. "Abderachman," I continued, "you are the only man who has spoken the truth. Go to Mecca! and may God protect yon on the journey! I should not wish to prevent you from performing your duty as a Mahometan."

Never were people more dumbfounded with surprise. They retreated, and formed a knot in consultation, and in about ten minutes they returned to me, old Moosa and Hadji Ali both leading the pilgrim Abderachman by the hands. They had given in; and Abderachman, the buffalo of the party, thanked me for my permission, and with tears in his eyes, as the camels were about to start, he at once said good-by. "Embrace him!" cried old Moosa and Hadji Ali; and in an instant, as I had formerly succumbed to the maid Barrake, I was actually kissed by the thick lips of Abderachman the unwashed! Poor fellow! this was sincere gratitude without the slightest humbug; therefore, although he was an odoriferous savage, I could not help shaking him by the hand and wishing him a prosperous journey, assuring him that I would watch over his comrades like a father, while in my service. In a few instants these curious people were led by a sudden and new impulse; my farewell had perfectly delighted old Moosa and Hadji Ali, whose hearts were won. "Say good-by to the Sit!" (the lady) they shouted to Abderachman; but I assured them that it was not necessary to go through the whole operation to which I had been subjected, and that she would be contented if he only kissed her hand. This he did with the natural grace of a savage, and was led away crying by his companions, who embraced him with tears, and they parted with the affection of brothers.

Now, to hard-hearted and civilized people, who often school themselves to feel nothing, or as little as they can, for anybody, it may appear absurd to say that the scene was affecting, but somehow or other it was. And in the course of half an hour, those who would have deserted had become stanch friends, and we were all, black and white, Mahometans and Christians, wishing the pilgrim God-speed upon his perilous journey to Mecca.

The camels started, and, if the scene was affecting, the invalids began to be more affected by the tartar emetic. This was the third act of the comedy. The plot had been thoroughly ventilated; the last act exhibited the perfect fidelity of my Tokrooris, in whom I subsequently reposed much confidence.

In the afternoon of that day the brothers Sherrif arrived. These were the most renowned of all the sword-hunters of the Hamrans, of whom I have already spoken. They were well mounted, and, having met our caravan of camels on the route, heavily laden with dried flesh, and thus seen proofs of our success, they now offered to join our party. I am sorry to be obliged to confess that my ally, Abou Do, although a perfect Nimrod in sport, an Apollo in personal appearance, and a gentleman in manner, was a mean, covetous, and grasping fellow, and withal absurdly jealous. Taher Sherrif was a more celebrated hunter, having had the experience of at least twenty years in excess of Abou Do; and although the latter was as brave and dexterous as Taher and his brothers, he wanted the cool judgment that is essential to a first-rate sportsman.

The following day was the new year, January 1st, 1862; and with the four brothers Sherrif and our party we formed a powerful body of hunters: six aggageers and myself all well mounted. With four gun-bearers and two camels, both of which carried water, we started in search of elephants.

Florian was unwell, and remained in camp.

The immediate neighborhood was a perfect exhibition of gun-arabic-bearing mimosas. At this season the gum was in perfection, and the finest quality was now before us in beautiful amber-colored masses upon the stems and branches, varying from the size of a nutmeg to that of an orange. So great was the quantity, and so excellent were the specimens, that, leaving our horses tied to trees, both the Arabs and myself gathered a large collection. This gum, although as hard as ice on the exterior, was limpid in the centre, resembling melted amber, and as clear as though refined by some artificial process. The trees were perfectly denuded of leaves from the extreme drought, and the beautiful balls of frosted yellow gum recalled the idea of the precious jewels upon the trees in the garden of the wonderful lamp of the "Arabian Nights." This gum was exceedingly sweet and pleasant to the taste; but, although of the most valuable quality, there was no hand to gather it in this forsaken although beautiful country; it either dissolved during the rainy season or was consumed by the baboons and antelopes. The aggageers took off from their saddles the skins of tanned antelope leather that formed the only covering to the wooden seats, and with these they made bundles of gum. When we remounted, every man was well laden.

We were thus leisurely returning home through alternate plains and low open forest of mimosa, when Taher Sherrif, who was leading the party, suddenly reined up his horse and pointed to a thick bush, beneath which was a large gray but shapeless mass. He whispered, as I drew near, "Oom gurrin" (mother of the horn), their name for the rhinoceros. I immediately dismounted, and with the short No. 10 Tatham rifle I advanced as near as I could, followed by Suleiman, as I had sent all my gum-bearers directly home by the river when we had commenced our circuit. As I drew near I discovered two rhinoceroses asleep beneath a thick mass of bushes. They were lying like pigs, close together, so that at a distance I had been unable to distinguish any exact form. It was an awkward place. If I were to take the wind fairly I should have to fire through the thick bush, which would be useless; therefore I was compelled to advance with the wind directly from me to them. The aggageers remained about a hundred yards distant, while I told Suleiman to return and hold my horse in readiness with his own. I then walked quietly to within about thirty yards of the rhinoceroses; but so curiously were they lying that it was useless to attempt a shot. In their happy dreams they must have been suddenly disturbed by the scent of an enemy, for, without the least warning, they suddenly sprang to their feet with astonishing quickness, and with a loud and sharp whiff, whiff, whiff! one of them charged straight at me. I fired my right-hand barrel in his throat, as it was useless to aim at the head protected by two horns at the nose. This turned him, but had no other effect, and the two animals thundered off together at a tremendous pace.

Now for a "tally-ho!" Our stock of gum was scattered on the ground, and away went the aggageers in full speed after the two rhinoceroses. Without waiting to reload, I quickly remounted my horse Tetel, and with Suleiman in company I spurred hard to overtake the flying Arabs. Tetel was a good strong cob, but not very fast; however, I believe he never went so well as upon that day, for, although an Abyssinian Horse, I had a pair of English spurs, which worked like missionaries. The ground was awkward for riding at full speed, as it was an open forest of mimosas, which, although wide apart, were very difficult to avoid, owing to the low crowns of spreading branches, and these, being armed with fish-hook thorns, would have been serious in a collision. I kept the party in view until in about a mile we arrived upon open ground. Here I again applied the spurs, and by degrees I crept up, always gaining, until I at length joined the aggageers.

Here was a sight to drive a hunter wild! The two rhinoceroses were running neck and neck, like a pair of horses in harness, but bounding along at tremendous speed within ten yards of the leading Hamran. This was Taher Sherrif, who, with his sword drawn and his long hair flying wildly behind him, urged his horse forward in the race, amid a cloud of dust raised by the two huge but active beasts, that tried every sinew of the horses. Roder Sherrif, with the withered arm, was second; with the reins hung upon the hawk-like claw that was all that remained of a hand, but with his naked sword grasped in his right, he kept close to his brother, ready to second his blow. Abou Do was third, his hair flying in the wind, his heels dashing against the flanks of his horse, to which he shouted in his excitement to urge him to the front, while he leaned forward with his long sword, in the wild energy of the moment, as though hoping to reach the game against all possibility.

Now for the spurs! and as these, vigorously applied, screwed an extra stride out of Tetel, I soon found myself in the ruck of men, horses, and drawn swords. There were seven of us, and passing Abou Do, whose face wore an expression of agony at finding that his horse was failing, I quickly obtained a place between the two brothers, Taher and Roder Sherrif. There had been a jealousy between the two parties of aggageers, and each was striving to outdo the other; thus Abou Do was driven almost to madness at the superiority of Taher's horse, while the latter, who was the renowned hunter of the tribe, was determined that his sword should be the first to taste blood. I tried to pass the rhinoceros on my left, so as to fire close into the shoulder my remaining barrel with my right hand, but it was impossible to overtake the animals, who bounded along with undiminished speed. With the greatest exertion of men and horses we could only retain our position within about three or four yards of their tails--just out of reach of the swords. The only chance in the race was to hold the pace until the rhinoceroses should begin. to flag. The horses were pressed to the utmost; but we had already run about two miles, and the game showed no signs of giving in. On they flew, sometimes over open ground, then through low bush, which tried the horses severely, then through strips of open forest, until at length the party began to tail off, and only a select few kept their places. We arrived at the summit of a ridge, from which the ground sloped in a gentle inclination for about a mile toward the river. At the foot of this incline was thick thorny nabbuk jungle, for which impenetrable covert the rhinoceroses pressed at their utmost speed.

Never was there better ground for the finish of a race. The earth was sandy, but firm, and as we saw the winning-post in the jungle that must terminate the hunt, we redoubled our exertions to close with the unflagging game. Suleiman's horse gave in--we had been for about twenty minutes at a killing pace. Tetel, although not a fast horse, was good for a distance, and he now proved his power of endurance, as I was riding at least two stone heavier than any of the party. Only four of the seven remained; and we swept down the incline, Taher Sherif still leading, and Abou Do the last! His horse was done, but not the rider; for, springing to the ground while at full speed, sword in hand, he forsook his tired horse, and, preferring his own legs, he ran like an antelope, and, for the first hundred yards I thought lie would really pass us and win the honor of first blow. It was of no use, the pace was too severe, and, although running wonderfully, he was obliged to give way to the horses. Only three now followed the rhinoceroses --Taher Sherrif, his brother Roder, and myself. I had been obliged to give the second place to Roder, as he was a mere monkey in weight; but I was a close third.

The excitement was intense. We neared the jungle, and the rhinoceroses began to show signs of flagging, as the dust puffed up before their nostrils, and, with noses close to the ground, they snorted as they still galloped on. Oh for a fresh horse! "A horse ! a horse! my kingdom for a horse!" We were within two hundred yards of the jungle; but the horses were all done. Tetel reeled as I urged him forward. Roder pushed ahead. We were close to the dense thorns, and the rhinoceroses broke into a trot; they were done! "Now, Taher, for-r-a-a-r-r-d! for-r-r-a-a-r-d, Taher!!"

Away he went. He was close to the very heels of the beasts, but his horse could do no more than his present pace; still he gained upon the nearest. He leaned forward with his sword raised for the blow. Another moment and the jungle would be reached! One effort more, and the sword flashed in the sunshine, as the rear-most rhinoceros disappeared in the thick screen of thorns, with a gash about a foot long upon his hind-quarters. Taher Sherrif shook his bloody sword in triumph above his head, but the rhinoceros was gone. We were fairly beaten, regularly outpaced; but I believe another two hundred yards would have given us the victory. "Bravo, Taher!" I shouted. He had ridden splendidly, and his blow had been marvellously delivered at an extremely long reach, as he was nearly out of his saddle when he sprang forward to enable the blade to obtain a cut at the last moment. He could not reach the hamstring, as his horse could not gain the proper position.

We all immediately dismounted. The horses were thoroughly done, and I at once loosened the girths and contemplated my steed Tetel, who, with head lowered and legs wide apart, was a tolerable example of the effects of pace. The other aggageers shortly arrived, and as the rival Abou Do joined us, Taher Sherrif quietly wiped the blood off his sword without making a remark. This was a bitter moment for the discomfited Abou Do.

There is only one species of rhinoceros in Abyssinia; this is the two-horned black rhinoceros, known in South Africa as the keitloa. This animal is generally five feet six inches to five feet eight inches high at the shoulder, and, although so bulky and heavily built, it is extremely active, as our long and fruitless hunt had shown us. The skin is about half the thickness of that of the hippopotamus, but of extreme toughness and closeness of texture. When dried and polished it resembles horn. Unlike the Indian species of rhinoceros, the black variety of Africa is free from folds, and the hide fits smoothly on the body like that of the buffalo. This two-horned black species is exceedingly vicious. It is one of the very few animals that will generally assume the offensive; it considers all creatures to be enemies, and, although it is not acute in either sight or hearing, it possesses so wonderful a power of scent that it will detect a stranger at a distance of five or six hundred yards should the wind be favorable.

Florian was now quite incapable of hunting, as he was in a weak state of health, and had for some months been suffering from chronic dysentery. I had several times cured him, but he had a weakness for the strongest black coffee, which, instead of drinking, like the natives, in minute cups, he swallowed wholesale in large basins several times a day; this was actual poison with his complaint, and he was completely ruined in health. At this time his old companion, Johann Schmidt, the carpenter, arrived, having undertaken a contract to provide for the Italian Zoological Gardens a number of animals. I therefore proposed that the two old friends should continue together, while I would hunt by myself, with the aggageers, toward the east and south. This arrangement was agreed to, and we parted.

Our camels returned from Geera with corn, accompanied by an Abyssinian hunter, who was declared by Abou Do to be a good man and dexterous with the sword. We accordingly moved our camp, said adieu to Florian and Johann, and penetrated still deeper into the country of the Bas-e.

Our course lay, as usual, along the banks of the river. We decided to encamp at a spot known to the Arabs as Deladilla. This was the forest upon the margin of the river where I had first shot the bull elephant when the aggageers fought with him upon foot. I resolved to fire the entire country on the following day, and to push still farther up the course of the Settite to the foot of the mountains, and to return to this camp in about a fortnight, by which time the animals that had been scared away by the fire would have returned. Accordingly, on the following morning, accompanied by a few of the aggageers, I started upon the south bank of the river, and rode for some distance into the interior, to the ground that was entirely covered with high withered grass. We were passing through a mass of kittar and thorn-bush, almost hidden by the immensely high grass, when, as I was ahead of the party, I came suddenly upon the tracks of rhinoceroses. These were so unmistakably recent that I felt sure we were not far from the animals themselves. As I had wished to fire the grass, I was accompanied by my Tokrooris and my horse-keeper, Mahomet No. 2. It was difficult ground for the men, and still more unfavorable for the horses, as large disjointed masses of stone were concealed in the high grass.

We were just speculating as to the position of the rhinoceros, and thinking how uncommonly unpleasant it would be should he obtain our wind, when whiff! whiff! whiff! We heard the sharp whistling snort, with a tremendous rush through the high grass and thorns close to us, and at the same moment two of these determined brutes were upon us in full charge. I never saw such a scrimmage. SAUVE QUI PEUT! There was no time for more than one look behind. I dug the spurs into Aggahr's flanks, and clasping him round the neck I ducked my head down to his shoulder, well protected with my strong hunting-cap, and kept the spurs going as hard as I could ply them, blindly trusting to Providence and my good horse. Over big rocks, fallen trees, thick kittar thorns, and grass ten feet high, with the two infernal animals in full chase only a few feet behind me! I heard their abominable whiffing close to me, but so did my good horse, and the good old hunter flew over obstacles in a way I should have thought impossible, and he dashed straight under the hooked thorn-bushes and doubled like a hare. The aggageers were all scattered; Mahomet No. 2 was knocked over by a rhinoceros; all the men were sprawling upon the rocks with their guns, and the party was entirely discomfited.

Having passed the kittar thorn I turned, and, seeing that the beasts had gone straight on, I brought Aggahr's head round and tried to give chase; but it was perfectly impossible. It was only a wonder that the horse had escaped in ground so difficult for riding. Although my clothes were of the strongest and coarsest Arab cotton cloth, which seldom tore, but simply lost a thread when caught in a thorn, I was nearly naked. My blouse was reduced to shreds. As I wore sleeves only half way from the shoulder to the elbow, my naked arms were streaming with blood. Fortunately my hunting-cap was secured with a chin strap, and still more fortunately I had grasped the horse's neck; otherwise I must have been dragged out of the saddle by the hooked thorns. All the men were cut and bruised, some having fallen upon their heads among the rocks, and others had hurt their legs in falling in their endeavors to escape. Mahomet No. 2, the horse-keeper, was more frightened than hurt, as he had been knocked down by the shoulder and not by the horn of the rhinoceros, as the animal had not noticed him; its attention was absorbed by the horse.

I determined to set fire to the whole country immediately, and descending the hill toward the river to obtain a favorable wind, I put my men in a line, extending over about a mile along the river's bed, and they fired the grass in different places. With a loud roar the flame leaped high in air and rushed forward with astonishing velocity. The grass was as inflammable as tinder, and the strong north wind drove the long line of fire spreading in every direction through the country.

CHAPTER X.

A day with the howartis--A hippo's gallant fight--Abou Do leaves us--Three yards from a lion--Days of delight--A lion's furious rage--Astounding courage of a horse.

A LITTLE before sunrise I accompanied the howartis, or hippopotamus-hunters, for a day's sport. At length we arrived at a large pool in which were several sand-banks covered with rushes, and many rocky islands. Among these rocks was a herd of hippopotami, consisting of an old bull and several cows. A young hippo was standing, like an ugly little statue, on a protruding rock, while another infant stood upon its mother's back that listlessly floated on the water.

This was an admirable place for the hunters. They desired me to lie down, and they crept into the jungle out of view of the river. I presently observed them stealthily descending the dry bed about two hundred paces above the spot where the hippos were basking behind the rocks. They entered the river and swam down the centre of the stream toward the rock. This was highly exciting. The hippos were quite unconscious of the approaching danger, as, steadily and rapidly, the hunters floated down the strong current. They neared the rock, and both heads disappeared as they purposely sank out of view; in a few seconds later they reappeared at the edge of the rock upon which the young hippo stood. It would be difficult to say which started first, the astonished young hippo into the water, or the harpoons from the hands of the howartis! It was the affair of a moment. The hunters dived as soon as they had hurled their harpoons, and, swimming for some distance under water, they came to the surface, and hastened to the shore lest an infuriated hippopotamus should follow them. One harpoon had missed; the other had fixed the bull of the herd, at which it had been surely aimed. This was grand sport! The bull was in the greatest fury, and rose to the surface, snorting and blowing in his impotent rage; but as the ambatch float was exceedingly large, and this naturally accompanied his movements, he tried to escape from his imaginary persecutor, and dived constantly, only to find his pertinacious attendant close to him upon regaining the surface. This was not to last long; the howartis were in earnest, and they at once called their party, who, with two of the aggageers, Abou Do and Suleiman, were near at hand. These men arrived with the long ropes that form a portion of the outfit of hippo hunting.

The whole party now halted on the edge of the river, while two men swam across with one end of the long rope. Upon gaining the opposite bank, I observed that a second rope was made fast to the middle of the main line. Thus upon our side we held the ends of two ropes, while on the opposite side they had only one; accordingly, the point of junction of the two ropes in the centre formed an acute angle. The object of this was soon practically explained. Two men upon our side now each held a rope, and one of these walked about ten yards before the other. Upon both sides of the river the people now advanced, dragging the rope on the surface of the water until they reached the ambatch float that was swimming to and fro, according to the movements of the hippopotamus below. By a dexterous jerk of the main line the float was now placed between the two ropes, and it was immediately secured in the acute angle by bringing together the ends of these ropes on our side.

The men on the opposite bank now dropped their line, and our men hauled in upon the ambatch float that was held fast between the ropes. Thus cleverly made sure, we quickly brought a strain upon the hippo, and, although I have had some experience in handling big fish, I never knew one to pull so lustily as the amphibious animal that we now alternately coaxed and bullied. He sprang out of the water, gnashed his huge jaws, snorted with tremendous rage, and lashed the river into foam. He then dived, and foolishly approached us beneath the water. We quickly gathered in the slack line, and took a round turn upon a large rock, within a few feet of the river. The hippo now rose to the surface, about ten yards from the hunters, and, jumping half out of the water, he snapped his great jaws together, endeavoring to catch the rope; but at the same instant two harpoons were launched into his side. Disdaining retreat, and maddened with rage, the furious animal charged from the depths of the river, and, gaining a footing, he reared his bulky form from the surface, came boldly upon the sand-bank, and attacked the hunters open-mouthed.

He little knew his enemy. They were not the men to fear a pair of gaping jaws, armed with a deadly array of tusks; but half a dozen lances were hurled at him, some entering his mouth from a distance of five or six paces. At the same time several men threw handfuls of sand into his enormous eyes. This baffled him more than the lances; he crunched the shafts between his powerful jaws like straws, but he was beaten by the sand, and, shaking his huge head, he retreated to the river. During his sally upon the shore two of the hunters had secured the ropes of the harpoons that had been fastened in his body just before his charge. He was now fixed by three of these deadly instruments; but suddenly one rope gave way, having been bitten through by the enraged beast, who was still beneath the water. Immediately after this he appeared on the surface, and, without a moment's hesitation, he once more charged furiously from the water straight at the hunters, with his huge mouth open to such an extent that he could have accommodated two inside passengers. Suleiman was wild with delight, and springing forward lance in hand, he drove it against the head of the formidable animal, but without effect. At the same time Abou Do met the hippo sword in hand, reminding me of Perseus slaying the sea-monster that would devour Andromeda; but the sword made a harmless gash, and the lance, already blunted against the rocks, refused to penetrate the tough hide. Once more handfuls of sand were pelted upon his face, and, again repulsed by this blinding attack, he was forced to retire to his deep hole and wash it from his eyes.

Six times during the fight the valiant bull hippo quitted his watery fortress and charged resolutely at his pursuers. He had broken several of their lances in his jaws, other lances had been hurled, and, falling upon the rocks, they were blunted and would not penetrate. The fight had continued for three hours, and the sun was about to set; accordingly the hunters begged me to give him the COUP DE GRACE, as they had hauled him close to the shore, and they feared he would sever the rope with his teeth. I waited for a good opportunity, when he boldly raised his head from water about three yards from the rifle, and a bullet from the little Fletcher between the eyes closed the last act. This spot was not far from the pyramidical hill beneath which I had fixed our camp, to which I returned after an amusing day's sport.

The next morning I started to the mountains to explore the limit that I had proposed for my expedition on the Settite. The Arabs had informed me that a river of some importance descended from the mountains and joined the main stream about twelve miles from our camp. In about three hours and a half we arrived at Hor Mehetape, the stream that the Arabs had reported. Although a powerful torrent during the rains, it was insignificant as one of the tributaries to the Settite, as the breadth did not exceed twenty-five yards. At this season it was nearly dry, and at no time did it appear to exceed the depth of ten or twelve feet. It was merely a rapid mountain torrent. But we were now among the mountains whose drainage causes the sudden rise of the Atbara and the Nile.

Abou Do and Suleiman had lately given us some trouble, especially the former, whose covetous nature had induced him to take much more than his share of the hides of rhinoceros and other animals shot. The horses of the aggageers had, moreover, been lamed by reckless riding, and Abou Do coolly proposed that I should lend them horses. Having a long journey before me, I refused, and they became discontented. It was time to part, and I ordered him and his people to return to Geera. As Taher Sherrif's party had disagreed with Abou Do some time previously, and had left us, we were now left without aggageers.

On the following day I succeeded in killing a buffalo, which I ordered my men, after they had flayed it, to leave as a bait for lions.

That night we were serenaded by the roaring of these animals in all directions, one of them having visited our camp, around which we discovered his footprints on the following morning. I accordingly took Taher Noor, with Hadji Ali and Hassan, two of my trusty Tokrooris, and went straight to the spot where I had left the carcass of the buffalo. As I had expected, nothing remained--not even a bone. The ground was much trampled, and tracks of lions were upon the sand; but the body of the buffalo had been dragged into the thorny jungle. I was determined, if possible, to get a shot; therefore I followed carefully the track left by the carcass, which had formed a path in the withered grass. Unfortunately the lions had dragged the buffalo down wind; therefore, after I had arrived within the thick nabbuk and high grass, I came to the conclusion that my only chance would be to make a long circuit, and to creep up wind through the thorns, until I should be advised by my nose of the position of the carcass, as it would by this time be in a state of putrefaction, and the lions would most probably be with the body. Accordingly I struck off to my left, and continuing straight forward for some hundred yards, I again struck into the thick jungle and came round to the wind. Success depended on extreme caution; therefore I advised my three men to keep close behind me with the spare rifles, as I carried my single-barrelled Beattie. This rifle was extremely accurate, therefore I had chosen it for this close work, when I expected to get a shot at the eye or forehead of a lion crouching in the bush.

Softly and with difficulty I crept forward, followed closely by my men, through the high withered grass, beneath the dense green nabbuk bushes, peering through the thick covert, with the nerves braced up to full pitch, and the finger on the trigger ready for any emergency. We had thus advanced for about half an hour, during which I frequently applied my nose to within a foot of the ground to catch the scent, when a sudden puff of wind brought the unmistakable smell of decomposing flesh. For the moment I halted, and, looking round to my men, I made a sign that we were near to the carcass, and that they were to be ready with the rifles. Again I crept gently forward, bending and sometimes crawling beneath the thorns to avoid the slightest noise. As I approached the scent became stronger, until I at length felt that I must be close to the cause.

This was highly exciting. Fully prepared for a quick shot, I stealthily crept on. A tremendous roar in the dense thorns within a few feet of me suddenly brought my rifle to the shoulder. Almost in the same instant I observed the three-quarter figure of either a lion or a lioness within three yards of me, on the other side of the bush under which I had been creeping. The foliage concealed the head, but I could almost have touched the shoulder with my rifle. Much depended upon the bullet, and I fired exactly through the shoulder. Another tremendous roar! and a crash in the bushes as the animal made a bound forward was succeeded immediately by a similar roar, as another lion took the exact position of the last, and stood wondering at the report of the rifle, and seeking for the cause of the intrusion. This was a grand lion with a shaggy mane; but my rifle was unloaded, and, keeping my eyes fixed on the beast, I stretched my hand back for a spare rifle. The lion remained standing, but gazing up wind with his head raised, snuffing in the air for a scent of the enemy. No rifle was put in my hand. I looked back for an instant, and saw my Tokrooris faltering about five yards behind me. I looked daggers at them, gnashing my teeth and shaking my fist. They saw the lion, and Taher Noor snatching a rifle from Hadji Ali was just about to bring it; when Hassan, ashamed, ran forward. The lion disappeared at the same moment. Never was such a fine chance lost through the indecision of the gun bearers! I made a vow never to carry a single-barrelled rifle again when hunting large game. If I had had my dear little Fletcher 24 I should have nailed the lion to a certainty.

However, there was not much time for reflection. Where was the first lion? Some remains of the buffalo lay upon my right, and I expected to find the lion most probably crouching in the thorns somewhere near us. Having reloaded, I took one of my Reilly No. 10 rifles and listened attentively for a sound. Presently I heard within a few yards a low growl. Taher Noor drew his sword and, with his shield before him, he searched for the lion, while I crept forward toward the sound, which was again repeated. A low roar, accompanied by a rush in the jungle, showed us a glimpse of the lion as he bounded off within ten or twelve yards; but I had no chance to fire. Again the low growl was repeated, and upon quietly creeping toward the spot I saw a splendid animal crouched upon the ground amid the withered and broken grass. The lioness lay dying with the bullet wound in the shoulder. Occasionally in her rage she bit her own paw violently, and then struck and clawed the ground. A pool of blood lay by her side. She was about ten yards from us, and I instructed my men to throw a clod of earth at her (there were no stones), to prove whether she could rise, while I stood ready with the rifle. She merely replied with a dull roar, and I terminated her misery by a ball through the head. She was a beautiful animal. The patch of the bullet was sticking in the wound. She was shot through both shoulders, and as we were not far from the tent I determined to have her brought to camp upon a camel as an offering to my wife. Accordingly I left my Tokrooris, while I went with Taher Noor to fetch a camel.

On our road through the thick jungle I was startled by a rush close to me. For the moment I thought it was a lion, but almost at the same instant I saw a fine nellut dashing away before me, and I killed it immediately with a bullet through the back of the neck. This was great luck, and we now required two camels, as in two shots I had killed a lioness and a nellut (A. Strepsiceros).

We remained for some time at our delightful camp at Delladilla. Every day, from sunrise to sunset, I was either on foot or in the saddle, without rest, except upon Sundays. As our camp was full of meat, either dried or in the process of drying in festoons upon the trees, we had been a great attraction to the beasts of prey, which constantly prowled around our thorn fence during the night. One night in particular a lion attempted to enter, but had been repulsed by the Tokrooris, who pelted him with firebrands. My people woke me up and begged me to shoot him; but as it was perfectly impossible to fire correctly through the hedge of thorns, I refused to be disturbed, but promised to hunt for him on the following day. Throughout the entire night the lion prowled around the camp, growling and uttering his peculiar guttural sigh. Not one of my people slept, as they declared he would bound into the camp and take somebody unless they kept up the watch-fires and drove him away with brands. The next day before sunrise I called Hassan and Hadji Ali, whom I lectured severely upon their cowardice on a former occasion, and received their promise to follow me to death. I intrusted them with my two Reillys No. 10, and with my little Fletcher in hand I determined to spend the whole day in searching every thicket of the forest for lions, as I felt convinced that the animal that had disturbed us during the night was concealed somewhere within the neighboring jungle.

The whole day passed fruitlessly. I had crept through the thickest thorns in vain; having abundance of meat, I had refused the most tempting shots at buffaloes and large antelopes, as I had devoted myself exclusively to lions. I was much disappointed, as the evening had arrived without a shot having been fired, and as the sun had nearly set I wandered slowly toward home. Passing through alternate open glades of a few yards' width, hemmed in on all sides by thick jungle, I was carelessly carrying my rifle upon my shoulder, as I pushed my way through the opposing thorns, when a sudden roar, just before me, at once brought the rifle upon full cock, and I saw a magnificent lion standing in the middle of the glade, about ten yards from me. He had been lying on the ground, and had started to his feet upon hearing me approach through the jungle. For an instant he stood in an attitude of attention, as we were hardly visible; but at the same moment I took a quick but sure shot with the little Fletcher. He gave a convulsive bound, but rolled over backward; before he could recover himself I fired the left-hand barrel.

It was a glorious sight. I had advanced a few steps into the glade, and Hassan had quickly handed me a spare rifle, while Taher Noor stood by me sword in hand. The lion in the greatest fury, with his shaggy mane bristling in the air, roared with death-like growls, as open-mouthed he endeavored to charge upon us; but he dragged his hind-quarters upon the ground, and I saw immediately that the little Fletcher had broken his spine. In his tremendous exertions to attack he rolled over and over, gnashing his horrible jaws and tearing holes in the sandy ground at each blow of his tremendous paws that would have crushed a man's skull like an egg-shell. Seeing that he was hors de combat I took it coolly, as it was already dusk, and the lion having rolled into a dark and thick bush I thought it would be advisable to defer the final attack, as he would be dead before morning. We were not ten minutes' walk from the camp, at which we quickly arrived, and my men greatly rejoiced at the discomfiture of their enemy, as they were convinced that he was the same lion that had attempted to enter the zareeba.

On the following morning before sunrise I started with nearly all my people and a powerful camel, with the intention of bringing the lion home entire. I rode my horse Tetel, who had frequently shown great courage, and I wished to prove whether he would advance to the body of a lion.

Upon arrival near the spot which we supposed to have been the scene of the encounter, we were rather puzzled, as there was nothing to distinguish the locality; one place exactly resembled another, as the country was flat and sandy, interspersed with thick jungle of green nabbuk. We accordingly spread out to beat for the lion. Presently Hadji Ali cried out, "There he lies, dead!" and I immediately rode to the spot together with the people. A tremendous roar greeted us as the lion started to his fore-feet, and with his beautiful mane erect and his great hazel eyes flashing fire he gave a succession of deep short roars, and challenged us to fight. This was a grand picture. He looked like a true lord of the forest; but I pitied the poor brute, as he was helpless, and although his spirit was game to the last, his strength was paralyzed by a broken back.

It was a glorious opportunity for the horse. At the first unexpected roar the camel had bolted with its rider. The horse had for a moment started on one side, and the men had scattered; but in an instant I had reined Tetel up, and I now rode straight toward the lion, who courted the encounter about twenty paces distant. I halted exactly opposite the noble-looking beast, who, seeing me in advance of the party, increased his rage and growled deeply, fixing his glance upon the horse. I now patted Tetel on the neck and spoke to him coaxingly. He gazed intently at the lion, erected his mane, and snorted, but showed no signs of retreat. "Bravo! old boy!" I said, and, encouraging him by caressing his neck with my hand, I touched his flank gently with my heel. I let him just feel my hand upon the rein, and with a "Come along, old lad," Tetel slowly but resolutely advanced step by step toward the infuriated lion, that greeted him with continued growls. The horse several times snorted loudly and stared fixedly at the terrible face before him; but as I constantly patted and coaxed him he did not refuse to advance. I checked him when within about six yards of the lion.

This would have made a magnificent picture, as the horse, with astounding courage, faced the lion at bay. Both animals kept their eyes fixed upon each other, the one beaming with rage, the other cool with determination. This was enough. I dropped the reins upon his neck; it was a signal that Tetel perfectly understood, and he stood firm as a rock, for he knew that I was about to fire. I took aim at the head of the glorious but distressed lion, and a bullet from the little Fletcher dropped him dead. Tetel never flinched at a shot. I now dismounted, and, having patted and coaxed the horse, I led him up to the body of the lion, which I also patted, and then gave my hand to the horse to smell. He snorted once or twice, and as I released my hold of the reins and left him entirely free, he slowly lowered his head and sniffed the mane of the dead lion. He then turned a few paces upon one side and commenced eating the withered grass beneath the nabbuk bushes.

My Arabs were perfectly delighted with this extraordinary instance of courage exhibited by the horse. I had known that the beast was disabled, but Tetel had advanced boldly toward the angry jaws of a lion that appeared about to spring. The camel was now brought to the spot and blindfolded, while we endeavored to secure the lion upon its back. As the camel knelt, it required the united exertions of eight men, including myself, to raise the ponderous animal and to secure it across the saddle.

Although so active and cat-like in its movements, a full-grown lion weighs about five hundred and fifty pounds. Having secured it we shortly arrived in camp. The COUP D'OEIL was beautiful, as the camel entered the enclosure with the shaggy head and massive paws of the dead lion hanging upon one flank, while the tail nearly descended to the ground upon the opposite side. It was laid at full length before my wife, to whom the claws were dedicated as a trophy to be worn around the neck as a talisman. Not only are the claws prized by the Arabs, but the mustache of the lion is carefully preserved and sewn in a leather envelope, to be worn as an amulet; such a charm is supposed to protect the wearer from the attacks of wild animals.

We were now destined to be deprived of two members of the party. Mahomet had become simply unbearable, and he was so impertinent that I was obliged to take a thin cane from one of the Arabs and administer a little physical advice. An evil spirit possessed the man, and he bolted off with some of the camel men who were returning to Geera with dried meat.

Our great loss was Barrake. She had persisted in eating the fruit of the hegleek, although she had suffered from dysentery upon several occasions. She was at length attacked with congestion of the liver. My wife took the greatest care of her, and for weeks she had given her the entire produce of the goats, hoping that milk would keep up her strength; but she died after great suffering, and we buried the poor creature, and moved our camp.

CHAPTER XI.

The bull-elephant--Daring Hamrans--The elephant helpless--Visited by a minstrel--A determined musician--The nest of the outlaws --The Atbara River

Having explored the Settite into the gorge of the mountain chain of Abyssinia, we turned due south from our camp at Deladilla, and at a distance of twelve miles reached the river Royan. Our course now was directed up this stream, and at the junction of the Hor Mai Gubba, or Habbuk River, some of my Arabs, observing fresh tracks of horses on the sand, went in search of the aggageers of Taher Sherrif's party, whom they had expected to meet at this point. Soon after, they returned with the aggageers, whose camp was but a quarter of a mile distant. I agreed to have a hunt for elephants the next day with Taher Sherrif, and before the following sunrise we had started up the course of the Royan for a favorite resort of elephants.

We had ridden about thirty miles, and were beginning to despair, when suddenly we turned a sharp angle in the watercourse, and Taher Sherrif, who was leading, immediately reined in his horse and backed him toward the party. I followed his example, and we were at once concealed by the sharp bend of the river. He now whispered that a bull-elephant was drinking from a hole it had scooped in the sand, not far around the corner. Without the slightest confusion the hunters at once fell quietly into their respective places, Taher Sherrif leading, while I followed closely in the line, with my Tokrooris bringing up the rear; we were a party of seven horses.

Upon turning the corner we at once perceived the elephant, that was still drinking. It was a fine bull. The enormous ears were thrown forward, as the head was lowered in the act of drawing up the water through the trunk. These shaded the eyes, and with the wind favorable we advanced noiselessly upon the sand to within twenty yards before we were perceived. The elephant then threw up its head, and with the ears flapping forward it raised its trunk for an instant, and then slowly but easily ascended the steep bank and retreated. The aggageers now halted for about a minute to confer together, and then followed in their original order up the crumbled bank. We were now on most unfavorable ground; the fire that had cleared the country we had hitherto traversed had been stopped by the bed of the torrent. We were thus plunged at once into withered grass above our heads, unless we stood in the stirrups. The ground was strewn with fragments of rock, and altogether it was ill-adapted for riding.

However, Taher Sherrif broke into a trot, followed by the entire party, as the elephant was not in sight. We ascended a hill, and when near the summit we perceived the elephant about eighty yards ahead. It was looking behind during its retreat, by swinging its huge head from side to side, and upon seeing us approach it turned suddenly round and halted.

"Be ready, and take care of the rocks!" said Taher Sherrif, as I rode forward by his side. Hardly had he uttered these words of caution when the bull gave a vicious jerk with its head, and with a shrill scream charged down upon us with the greatest fury. Away we all went, helter-skelter, through the dry grass, which whistled in my ears, over the hidden rocks, at full gallop, with the elephant tearing after us for about a hundred and eighty yards at a tremendous pace. Tetel was a sure-footed horse, and being unshod he never slipped upon the stones. Thus, as we all scattered in different directions, the elephant became confused and relinquished the chase. It had been very near me at one time, and in such ground I was not sorry when it gave up the hunt. We now quickly united and again followed the elephant, that had once more retreated. Advancing at a canter, we shortly came in view. Upon seeing the horses the bull deliberately entered a stronghold composed of rocky and uneven ground, in the clefts of which grew thinly a few leafless trees of the thickness of a man's leg. It then turned boldly toward us, and stood determinedly at bay.

Now came the tug of war! Taher Sherrif came close to me, and said, "You had better shoot the elephant, as we shall have great difficulty in this rocky ground." This I declined, as I wished the fight ended as it had been commenced, with the sword; and I proposed that he should endeavor to drive the animal to more favorable ground. "Never mind," replied Taher, "Inshallah (please God) he shall not beat us." He now advised me to keep as close to him as possible and to look sharp for a charge.

The elephant stood facing us like a statue; it did not move a muscle beyond a quick and restless action of the eyes, that were watching all sides. Taher Sherrif and his youngest brother, Ibrahim, now separated, and each took opposite sides of the elephant, and then joined each other about twenty yards behind it. I accompanied them, until Taher advised me to keep about the same distance upon the left flank. My Tokrooris kept apart from the scene, as they were not required. In front of the elephant were two aggageers, one of whom was the renowned Roder Sherrif, with the withered arm. All being ready for action, Roder now rode slowly toward the head of the cunning old bull, who was quietly awaiting an opportunity to make certain of some one who might give him a good chance.

Roder Sherrif rode a bay mare that, having been thoroughly trained to these encounters, was perfect at her work. Slowly and coolly she advanced toward her wary antagonist until within about eight or nine yards of the elephant's head. The creature never moved, and the mise en scene was beautiful. Not a word was spoken, and we kept our places amid utter stillness, which was at length broken by a snort from the mare, who gazed intently at the elephant, as though watching for the moment of attack.

One more pace forward, and Roder sat coolly upon his mare, with his eyes fixed upon those of the elephant. For an instant I saw the white of the eye nearest to me. "Look out, Roder, he's coming!" I exclaimed. With a shrill scream the elephant dashed upon him like an avalanche.

Round went the mare as though upon a pivot, and away, over rocks and stones, flying like a gazelle, with the monkey-like form of little Roder Sherrif leaning forward, and looking over his left shoulder as the elephant rushed after him.

For a moment I thought he must be caught. Had the mare stumbled, all were lost; but she gained in the race after a few quick, bounding strides, and Roder, still looking behind him, kept his distance so close to the elephant that its outstretched trunk was within a few feet of the mare's tail.

Taher Sherrif and his brother Ibrahim swept down like falcons in the rear. In full speed they dexterously avoided the trees until they arrived upon open ground, when they dashed up close to the hind-quarters of the furious elephant, which, maddened with the excitement, heeded nothing but Roder and his mare, that were almost within its grasp. When close to the tail of the elephant Taher Sherrif's sword flashed from its sheath, as grasping his trusty blade he leaped nimbly to the ground, while Ibrahim caught the reins of his horse. Two or three bounds on foot, with the sword clutched in both hands, and he was close behind the elephant. A bright glance shone like lightning as the sun struck upon the descending steel; this was followed by a dull crack, as the sword cut through skin and sinews, and settled deep in the bone, about twelve inches above the foot. At the next stride the elephant halted dead short in the midst of its tremendous charge. Taher had jumped quickly on one side, and had vaulted into the saddle with his naked sword in hand. At the same moment Roder, who had led the chase, turned sharp round, and again faced the elephant as before. Stooping quickly from the saddle, he picked up from the ground a handful of dirt, which he threw into the face of the vicious-looking animal, that once more attempted to rush upon him. It was impossible! The foot was dislocated, and turned up in front like an old shoe. In an instant Taher was once more on foot, and the sharp sword slashed the remaining leg.

The great bull-elephant could not move! The first cut with the sword had utterly disabled it; the second was its deathblow. The arteries of the leg were divided, and the blood spouted in jets from the wounds. I wished to terminate its misery by a bullet behind the ear, but Taher Sherrif begged me not to fire, as the elephant would quickly bleed to death without pain, and an unnecessary shot might attract the Base, who would steal the flesh and ivory during our absence. We were obliged to return immediately to our far distant camp, and the hunters resolved to accompany their camels to the spot on the following day. We turned our horses' heads, and rode directly toward home, which we did not reach until nearly midnight, having ridden upward of sixty miles during the day.

The hunting of Taher Sherrif and his brothers was superlatively beautiful; with an immense amount of dash there was a cool, sportsman-like manner in their mode of attack that far excelled the impetuous and reckless onset of Abou Do. It was difficult to decide which to admire the more, the coolness and courage of him who led the elephant, or the extraordinary skill and activity of the aggahr who dealt the fatal blow.

After hunting and exploring for some days in this neighborhood, I determined to follow the bed of the Royan to its junction with the Settite. We started at daybreak, and after a long march along the sandy bed, hemmed in by high banks or by precipitous cliffs of sandstone, we arrived at the junction.

Having explored the entire country and enjoyed myself thoroughly, I was now determined to pay our promised visit to Mek Nimmur. Since our departure from the Egyptian territory his country had been invaded by a large force, according to orders sent from the Governor-General of the Soudan. Mek Nimmur as usual retreated to the mountains, but Mai Gubba and a number of his villages were utterly destroyed by the Egyptians. He would under these circumstances be doubly suspicious of strangers.

We were fortunate, however, in unexpectedly meeting a party of Mek Nimmur's followers on a foray, who consented to guide us to his encampment. Accordingly on March 20th, we found ourselves in a rich and park-like valley occupied by his people, and the day following was spent in receiving visits from the head men. Messengers soon after arrived from Mek

Nimmur inviting us to pay him a visit at his residence.

As we were conversing with Mek Nimmur's messengers through the medium of Taher Noor, who knew their language, our attention was attracted by the arrival of a tremendous swell, who at a distance I thought must be Mek Nimmur himself. A snow-white mule carried an equally snow-white person, whose tight white pantaloons looked as though he had forgotten his trousers and had mounted in his drawers. He carried a large umbrella to shade his complexion; a pair of handsome silver-mounted pistols were arranged upon his saddle, and a silver-hilted curved sword, of the peculiar Abyssinian form, hung by his side. This grand personage was followed by an attendant, also mounted upon a mule, while several men on foot accompanied them, one of whom carried his lance and shield. Upon near approach he immediately dismounted and advanced toward us, bowing in a most foppish manner, while his attendant followed him on foot with an enormous violin, which he immediately handed to him. This fiddle was very peculiar in shape, being a square, with an exceedingly long neck extending from one corner. Upon this was stretched a solitary string, and the bow was very short and much bent. This was an Abyssinian Paganini. He was a professional minstrel of the highest grade, who had been sent by Mek Nimmur to welcome us on our arrival.

These musicians are very similar to the minstrels of ancient times. They attend at public rejoicings, and at births, deaths, and marriages of great personages, upon which occasions they extemporize their songs according to circumstances. My hunting in the Base country formed his theme, and for at least an hour he sang of my deeds in an extremely loud and disagreeable voice, while he accompanied himself upon his fiddle, which he held downward like a violoncello. During the whole of his song he continued in movement, marching with a sliding step to the front, and gliding to the right and left in a manner that, though intended to be graceful, was extremely comic. The substance of this minstrelsy was explained to me by Taher Noor, who listened eagerly to the words, which he translated with evident satisfaction. Of course, like all minstrels, he was an absurd flatterer, and, having gathered a few facts for his theme, he wandered slightly from the truth in his poetical description of my deeds.

He sang of me as though I had been Richard Coeur de Lion, and recounted, before an admiring throng of listeners, how I had wandered with a young wife from my own distant country to fight the terrible Base; how I had slain them in a single combat, and bow elephants and lions were struck down like lambs and kids by my hands. That during my absence in the hunt my wife had been carried off by the Base; that I had, on my return to my pillaged camp, galloped off in chase, and, overtaking the enemy, hundreds had fallen by my rifle and sword, and I had liberated and recovered the lady, who now had arrived safe with her lord in the country of the great Mek Nimmur, etc., etc.

This was all very pretty, no doubt, and as true as most poetical and musical descriptions; but I felt certain that there must be something to pay for this flattering entertainment. If you are considered to be a great man, a PRESENT is invariably expected in proportion to your importance. I suggested to Taher Noor that I must give him a couple of dollars. "What!" said Taher Noor, "a couple of dollars? Impossible! a musician of his standing is accustomed to receive thirty and forty dollars from great people for so beautiful and honorable a song."

This was somewhat startling. I began to reflect upon the price of a box at Her Majesty's Theatre in London; but there I was not the hero of the opera. This minstrel combined the whole affair in a most simple manner. He was Verdi, Costa, and orchestra all in one. He was a thorough Macaulay as historian, therefore I had to pay the composer as well as the fiddler. I compromised the matter, and gave him a few dollars, as I understood that he was Mek Nimmur's private minstrel; but I never parted with my dear Maria Theresa (* The Austrian dollar, that is the only large current coin in that country.) with so much regret as upon that occasion, and I begged him not to incommode himself by paying us another visit, or, should he be obliged to do so, I trusted he would not think it necessary to bring his violin.

The minstrel retired in the same order that he had arrived, and I watched his retreating figure with unpleasant reflections, that were suggested by doubts as to whether I had paid him too little or too much. Taher Noor thought that he was underpaid; my own opinion was that I had brought a curse upon myself equal to a succession of London organ-grinders, as I fully expected that other minstrels, upon hearing of the Austrian dollars, would pay us a visit and sing of my great deeds.

In the afternoon we were sitting beneath the shade of our tamarind tree, when we thought we could perceive our musical friend returning. As he drew near, we were convinced that it was the identical minstrel, who had most probably been sent with a message from Mek Nimmur. There he was, in snow-white raiment, on the snow-white mule, with the mounted attendant and the violin as before. He dismounted upon arrival opposite the camp, and approached with his usual foppish bow; but we looked on in astonishment: it was not our Paganini, it was ANOTHER MINSTREL! who was determined to sing an ode in our praise. I felt that this was an indirect appeal to Maria Theresa, and I at once declared against music. I begged him not to sing; "my wife had a headache--I disliked the fiddle--could He play anything else instead?" and I expressed a variety of polite excuses, but to no purpose; he insisted upon singing. If I disliked the fiddle, he would sing without an accompaniment, but he could not think of insulting so great a man as myself by returning without an ode to commemorate our arrival.

I was determined that he should NOT sing; he was determined that he WOULD, therefore I desired him to leave my camp. This he agreed to do, provided I would allow him to cross the stream and sing to my Tokrooris in my praise, beneath a neighboring tree about fifty yards distant. He remounted his mule with his violin, to ford the muddy stream, and descended the steep bank, followed by his attendant on foot, who drove the unwilling mule. Upon arrival at the brink of the dirty brook, that was about three feet deep, the mule positively refused to enter the water, and stood firm with its fore feet sunk deep in the mud. The attendant attempted to push it on behind, and at the same time gave it a sharp blow with his sheathed sword. This changed the scene to the "opera comique." In one instant the mule gave so vigorous and unexpected a kick into the bowels of the attendant that he fell upon his back, heels, uppermost, while at the same moment the minstrel, in his snow-white garments, was precipitated head fore-most into the muddy brook, and, for the moment disappearing, the violin alone could be seen floating on the surface. A second later, a wretched-looking object, covered with slime and filth, emerged from the slongh; this was Paganini the second! who, after securing his fiddle, that had stranded on a mud-bank, scrambled up the steel slope, amid the roars of laughter of my people and of ourselves, while the perverse mule, having turned harmony into discord, kicked up its heels and galloped off, braying an ode in praise of liberty, as the "Lay of the Last Minstrel." The discomfited fiddler was wiped down by my Tokrooris, who occasionally burst into renewed fits of laughter during the operation. The mule was caught, and the minstrel remounted, and returned home completely out of tune.

On the following morning at sunrise I mounted my horse, and, accompanied by Taher Noor and Bacheet, I rode to pay my respects to Mek Nimmur. Our route lay parallel to the stream, and after a ride of about two miles through a fine park-like country, bounded by the Abyssinian Alps about fifteen miles distant, I observed a crowd of people round a large tamarind tree, near which were standing a number of horses, mules, and dromedaries. This was the spot upon which I was to meet Mek Nimmur. Upon my approach the crowd opened, and, having dismounted, I was introduced by Taher Noor to the great chief. He was a man of about fifty, and exceedingly dirty in appearance. He sat upon an angarep, surrounded by his people; lying on either side upon his seat were two brace of pistols, and within a few yards stood his horse ready saddled. He was prepared for fight or flight, as were also his ruffianly looking followers, who were composed of Abyssinians and Jaleens. After a long and satisfactory conversation I retired. Immediately on my arrival at camp I despatched Wat Gamma with a pair of beautiful double-barrelled pistols, which I begged Mek Nimmur to accept. On March 27th we said good-by and started for the Bahr Salaam.

The next few days we spent in exploring the Salaam and Angrab rivers. They are interesting examples of the destructive effect of water, that has during the course of ages cut through and hollowed out, in the solid rock, a succession of the most horrible precipices and caverns, in which the maddened torrents, rushing from the lofty chain of mountains, boil along until they meet the Atbara and assist to flood the Nile. No one could explore these tremendous torrents, the Settite, Royan, Angrab, Salaam, and Atbara, without at once comprehending their effect upon the waters of the Nile. The magnificent chain of mountains from which they flow is not a simple line of abrupt sides, but the precipitous slopes are the walls of a vast plateau, that receives a prodigious rainfall in June, July, August, and until the middle of September, the entire drainage of which is carried away by the above-named channels to inundate Lower Egypt.

I thoroughly explored the beautiful country of the Salaam and Angrab, and on the 14th of April we pushed on for Gallabat, the frontier market-town of

Abyssinia.

We arrived at our old friend, the Atbara River, at the sharp angle as it issues from the mountains. At this place it was in its infancy. The noble Atbara, whose course we had tracked for hundreds of weary miles, and whose tributaries we had so carefully examined, was here a second-class mountain torrent, about equal to the Royan, and not to be named in comparison with the Salaam or Angrab. The power of the Atbara depended entirely upon the western drainage of the Abyssinian Alps; of itself it was insignificant until aided by the great arteries of the mountain-chain. The junction of the Salaam at once changed its character, and the Settite or Taccazzy completed its importance as the great river of Abyssinia, that has washed down the fertile soil of those regions to create the Delta of Lower Egypt, and to perpetuate that Delta by annual deposits, that ARE NOW FORMING A NEW EGYPT BENEATH THE WATERS OF THE MEDITERRANEAN. We had seen the Atbara a bed of glaring sand--a mere continuation of the burning desert that surrounded its course--fringed by a belt of withered trees, like a monument sacred to the memory of a dead river. We had seen the sudden rush of waters when, in the still night, the mysterious stream had invaded the dry bed and swept all before it like an awakened giant; we knew at that moment "the rains were falling in Abyssinia," although the sky above us was without a cloud. We had subsequently witnessed that tremendous rainfall, and seen the Atbara at its grandest flood. We had traced each river and crossed each tiny stream that fed the mighty Atbara from the mountain-chain, and we now, after our long journey, forded the Atbara in its infancy, hardly knee-deep, over its rocky bed of about sixty yards' width, and camped in the little village of Toganai, on the rising ground upon the opposite side. It was evening, and we sat upon an angarep among the lovely hills that surrounded us, and looked down upon the Atbara for the last time, as the sun sank behind the rugged mountain of Ras el Feel (the elephant's head). Once more I thought of that wonderful river Nile, that could flow forever through the exhausting deserts of sand, while the Atbara, during the summer months, shrank to a dry skeleton, although the powerful affluents, the Salaam and the Settite, never ceased to flow; every drop of their waters was evaporated by the air and absorbed by the desert sand in the bed of the Atbara, two hundred miles above its junction with the Nile!

The Atbara exploration was completed, and I looked forward to the fresh enterprise of exploring new rivers and lower latitudes, that should unravel the mystery of the Nile!

CHAPTER XII.

Abyssinian slave-girls--Khartoum--The Soudan under Egyptian rule --Slave-trade in the Soudan--The obstacles ahead.

A rapid march of sixteen miles brought us to Metemma or Gallabat. As we descended the valley we perceived great crowds of people in and about the town, which, in appearance, was merely a repetition of Katariff. It was market-day, and as we descended the hill and arrived in the scene below, with our nine camels heavily laden with the heads and horns of a multitude of different beasts, from the gaping jaws of hippopotami to the vicious-looking heads of rhinoceroses and buffaloes, while the skins of lions and various antelopes were piled above masses of the much-prized hide of the rhinoceros, we were beset by crowds of people, who were curious to know whence so strange a party had come. We formed a regular procession through the market, our Tokrooris feeling quite at home among so many of their brethren.

While here I visited the establishments of the various slave merchants. These were arranged under large tents formed of matting, and contained many young girls of extreme beauty, ranging from nine to seventeen years of age. These lovely captives, of a rich brown tint, with delicately formed features, and eyes like those of the gazelle, were natives of the Galla, on the borders of Abyssinia, from which country they were brought by the Abyssinian traders to be sold for the Turkish harems. Although beautiful, these girls are useless for hard labor; they quickly fade away, and die unless kindly treated. They are the Venuses of that country, and not only are their faces and figures perfection, but they become extremely attached to those who show them kindness, and they make good and faithful wives. There is something peculiarly captivating in the natural grace and softness of these young beauties, whose hearts quickly respond to those warmer feelings of love that are seldom known among the sterner and coarser tribes. Their forms are peculiarly elegant and graceful; the hands and feet are exquisitely delicate; the nose is generally slightly aquiline, the nostrils large and finely shaped; the hair is black and glossy, reaching to about the middle of the back, but rather coarse in texture. These girls, although natives of Galla, Invariably call themselves Abyssinians, and are generally known under that name. They are exceedingly proud and high-spirited, and are remarkably quick at learning. At Khartoum several of the Europeans of high standing have married these charming ladies, who have invariably rewarded their husbands by great affection and devotion. The price of one of these beauties of nature at Gallabat was from twenty-five to forty dollars!

On the march from Gallabat to the Rahad River I was so unfortunate as to lose my two horses, Gazelle and Aggahr. The sudden change of food from dry grass to the young herbage which had appeared after a few showers, brought on inflammation of the bowels, which carried them off in a few hours. We now travelled for upward of a hundred miles along the bank of the Rahad, through a monotonous scene of flat alluvial soil. The entire country would be a Mine of wealth were it planted with cotton, Which could be transported by river to Katariff, and thence directly to Souakim.

I shall not weary the reader with the details of the rest of our journey to Khartoum, the capital of the Soudan provinces, at which we arrived on the 11th of June.

The difference between the appearance of Khartoum at the distance of a mile, with the sun shining upon the bright river Nile in the foreground, and its appearance upon close inspection, was equal to the difference in the scenery of a theatre as regarded from the boxes or from the stage. Even that painful exposure of an optical illusion would be trifling compared with the imposture of Khartoum. The sense of sight had been deceived by distance, but the sense of smell was outraged by innumerable nuisances, when we set foot within the filthy and miserable town. After winding through some narrow, dusty lanes, hemmed in by high walls of sun-baked bricks that had fallen in gaps in several places, exposing gardens of prickly pears and date palms, we at length arrived at a large open place, that, if possible, smelt more strongly than the landing spot. Around this square, which was full of holes where the mud had been excavated for brick-making, were the better class of houses; this was the Belgravia of Khartoum. In the centre of a long mud wall, ventilated by certain attempts at frameless windows, guarded by rough wooden bars, we perceived a large archway with closed doors. Above this entrance was a shield, with a device that gladdened my English eyes: there was the British lion and the unicorn! Not such a lion as I had been accustomed to meet in his native jungles, a yellow cowardly fellow that had often slunk away from the very prey from which I had driven him; but a real red British lion, that, although thin and ragged in the unhealthy climate of Khartoum, looked as though he was pluck to the back-bone.

This was the English Consulate. The consul was absent, in the hope of meeting Speke and Grant in the upper Nile regions, on the road from Zanzibar, but he had kindly placed rooms at our disposal.

For some months we resided at Khartoum, as it was necessary to make extensive preparations for the White Nile expedition, and to await the arrival of the north wind, which would enable us to start early in December. Although the north and south winds blow alternately for six months, and the former commences in October, it does not extend many degrees southward until the beginning of December. This is a great drawback to White Nile exploration, as, when near the north side of the equator, the dry season commences in November and closes in February; thus the departure from Khartoum should take place by a steamer in the latter part of September. That would enable the traveller to leave Gondokoro, lat. N. 4 "degrees" 54', shortly before November. He would then secure three months of favorable weather for an advance inland.

Khartoum is a wretchedly unhealthy town, containing about thirty thousand inhabitants, exclusive of troops. In spite of its unhealthiness and low situation, on a level with the river at the junction of the Blue and White Niles, it is the general emporium for the trade of the Soudan, from which the productions of the country are transported to Lower Egypt, i.e. ivory, hides, senna, gum arabic, and beeswax. During my experience of Khartoum it was the hotbed of the slave-trade. It will be remarked that the exports from the Soudan are all natural productions. There is nothing to exhibit the industry or capacity of the natives. The ivory is the produce of violence and robbery; the hides are the simple sun-dried skins of oxen; the senna grows wild upon the desert; the gum arabic exudes spontaneously from the bushes of the jungle; and the bees-wax is the produce of the only industrious creatures in that detestable country.

When we regard the general aspect of the Soudan, it is extreme wretchedness. The rainfall is uncertain and scanty; thus the country is a desert, dependent entirely upon irrigation. Although cultivation is simply impossible without a supply of water, one of the most onerous taxes is that upon the sageer or water-wheel, with which the fields are irrigated on the borders of the Nile. It would appear natural that, instead of a tax, a premium should be offered for the erection of such means of irrigation, which would increase the revenue by extending cultivation, the produce of which might bear an impost. With all the talent and industry of the native Egyptians, who must naturally depend upon the waters of the Nile for their existence, it is extraordinary that for thousands of years they have adhered to their original simple form of mechanical irrigation, without improvement.

The general aspect of the Soudan is that of misery; nor is there a single feature of attraction to recompense a European for the drawbacks of pestilential climate and brutal associations. To a stranger it appears a superlative folly that the Egyptian Government should have retained a possession the occupation of which is wholly unprofitable, the receipts being far below the expenditure malgre the increased taxation. At so great a distance from the sea-coast and hemmed in by immense deserts, there is a difficulty of transport that must nullify all commercial transactions on an extended scale.

The great and most important article of commerce as an export from the Soudan is gum arabic. This is produced by several species of mimosa, the finest quality being a product of Kordofan; the other natural productions exported are senna, hides, and ivory. All merchandise both to and from the Soudan must be transported upon camels, no other animals being adapted to the deserts. The cataracts of the Nile between Assouan and Khartoum rendering the navigation next to impossible, camels are the only medium of transport, and the uncertainty of procuring them without great delay is the trader's greatest difficulty. The entire country is subject to droughts that occasion a total desolation, and the want of pasture entails starvation upon both cattle and camels, rendering it at certain seasons impossible to transport the productions of the country, and thus stagnating all enterprise. Upon existing conditions the Soudan is worthless, having neither natural capabilities nor political importance; but there is, nevertheless, a reason that first prompted its occupation by the Egyptians, and that is, THE SOUDAN SUPPLIES SLAVES.

Without the White Nile trade Khartoum* would almost cease to exist; (* This was written about twenty years ago, and does not apply to the Khartoum of to-day. In 1869 The Khedive of Egypt despatched an expedition under Sir Samuel Baker to suppress slavery in the Soudan and Central Africa. To the success of that expedition, and to the efforts of Colonel (now General) Gordon, who succeeded to the command of the Soudan, was owing the suppression of the traffic in slaves. Within the last few weeks, under the stress of circumstances, General Gordon has been forced to promise the removal of this prohibition of slavery.--E. J. W.) and that trade is kidnapping and murder. The character of the Khartoumers needs no further comment. The amount of ivory brought down from the White Nile is a mere bagatelle as an export, the annual value being about 40,000 pounds.

The people for the most part enraged in the nefarious traffic of the White Nile are Syrians, Copts, Turks, Circassians, and some few EUROPEANS. So closely connected with the difficulties of my expedition is that accursed slave-trade, that the so-called ivory trade of the White Nile requires an explanation.

Throughout the Soudan money is exceedingly scarce and the rate of interest exorbitant, varying, according to the securities, from thirty-six to eighty per cent. This fact proves general poverty and dishonesty, and acts as a preventive to all improvement. So high and fatal a rate deters all honest enterprise, and the country must lie in ruin under such a system. The wild speculator borrows upon such terms, to rise suddenly like a rocket, or to fall like its exhausted stick. Thus, honest enterprise being impossible, dishonesty takes the lead, and a successful expedition to the White Nile is supposed to overcome all charges. There are two classes of White Nile traders, the one possessing capital, the other being penniless adventurers. The same system of operations is pursued by both, but that of the former will be evident from the description of the latter.

A man without means forms an expedition, and borrows money for this purpose at 100 per cent. after this fashion: he agrees to repay the lender in ivory at one-half its market value. Having obtained the required sum, he hires several vessels and engages from 100 to 300 men, composed of Arabs and runaway villains from distant countries, who have found an asylum from justice in the obscurity of Khartoum. He purchases guns and large quantities of ammunition for his men, together with a few hundred pounds of glass beads. The piratical expedition being complete, he pays his men five months' wages in advance, at the rate of forty-five piastres (nine shillings) per month, and he agrees to give them eighty piastres per month for any period exceeding the five months for which they are paid. His men receive their advance partly in cash and partly in cotton stuffs for clothes at an exorbitant price. Every man has a strip of paper, upon which is written, by the clerk of the expedition, the amount he has received both in goods and money, and this paper he must produce at the final settlement.

The vessels sail about December, and on arrival at the desired locality the party disembark and proceed into the interior, until they arrive at the village of some negro chief, with whom they establish an intimacy.

Charmed with his new friends, the power of whose weapons he acknowledges, the negro chief does not neglect the opportunity of seeking their alliance to attack a hostile neighbor. Marching throughout the night, guided by their negro hosts, they bivouac within an hour's march of the unsuspecting village doomed to an attack about half an hour before break of day. The time arrives, and, quietly surrounding the village while its occupants are still sleeping, they fire the grass huts in all directions and pour volleys of musketry through the flaming thatch. Panic-stricken, the unfortunate victims rush from their burning dwellings, and the men are shot down like pheasants in a battue, while the women and children, bewildered in the danger and confusion, are kidnapped and secured. The herds of cattle, still within their kraal or "zareeba," are easily disposed of, and are driven off with great rejoicing, as the prize of victory. The women and children are then fastened together, and the former secured in an instrument called a sheba, made of a forked pole, the neck of the prisoner fitting into the fork, and secured by a cross-piece lashed behind, while the wrists, brought together in advance of the body, are tied to the pole. The children are then fastened by their necks with a rope attached to the women, and thus form a living chain, in which order they are marched to the head-quarters in company with the captured herds.

This is the commencement of business. Should there be ivory in any of the huts not destroyed by the fire, it is appropriated. A general plunder takes place. The trader's party dig up the floors of the huts to search for iron hoes, which are generally thus concealed, as the greatest treasure of the negroes;

the granaries are overturned and wantonly destroyed, and the hands are cut off the bodies of the slain, the more easily to detach the copper or iron bracelets that are usually worn. With this booty the TRADERS return to their negro ally. They have thrashed and discomfited his enemy, which delights him; they present him with thirty or forty head of cattle, which intoxicates him with joy, and a present of a pretty little captive girl of about fourteen completes his happiness.

An attack or razzia, such as described, generally leads to a quarrel with the negro ally, who in his turn is murdered and plundered by the trader--his women and children naturally becoming slaves.

A good season for a party of a hundred and fifty men should produce about two hundred cantars (20,000 lbs.) of ivory, valued at Khartoum at 4,000 pounds. The men being paid in slaves, the wages should be NIL, and there should be a surplus of four or five hundred slaves for the trader's own profit--worth on an average five to six pounds each.

The amiable trader returns from the White Nile to Khartoum; hands over to his creditor sufficient ivory to liquidate the original loan of 1,000 pounds, and, already a man of capital, he commences as an independent trader.

Such was the White Nile trade when I prepared to start from Khartoum on my expedition to the Nile sources. Every one in Khartoum, with the exception of a few Europeans, was in favor of the slave-trade, and looked with jealous eyes upon a stranger venturing within the precincts of their holy land--a land sacred to slavery and to every abomination and villainy that man can commit.

The Turkish officials pretended to discountenance slavery; at the same time every house in Khartoum was full of slaves, and the Egyptian officers had been in the habit of receiving a portion of their pay in slaves, precisely as the men employed on tile White Nile were paid by their employers. The Egyptian authorities looked upon the exploration of the White Nile by a European traveller as an infringement of the slave territory that resulted from espionage, and every obstacle was thrown in my way.

To organize an enterprise so difficult that it had hitherto defeated the whole world, required a careful selection of attendants, and I looked with despair at the prospect before me. The only men procurable for escort were the miserable cut-throats of Khartoum, accustomed to murder and pillage in the White Nile trade, and excited not by the love of adventure, but by the desire for plunder. To start with such men appeared mere insanity.

There was a still greater difficulty in connection with the White Nile. For years the infernal traffic in slaves and its attendant horrors had existed like a pestilence in the negro countries, and had so exasperated the tribes that people who in former times were friendly had become hostile to all comers. An exploration to the Nile sources was thus a march through an enemy's country, and required a powerful force of well-armed men. For the traders there was no great difficulty, as they took the initiative in hostilities, and had fixed camps as "points d'appui;" but for an explorer there was no alternative, but he must make a direct forward march with no communications with the rear. I had but slight hope of success without assistance from the authorities in the shape of men accustomed to discipline. I accordingly wrote to the British consul at Alexandria, and requested him to apply for a few soldiers and boats to aid me in so difficult an enterprise. After some months' delay, owing to the great distance from Khartoum, I received a reply inclosing a letter from Ismail Pacha (the present Viceroy), the regent during the absence of Said Pacha, REFUSING the application.

I confess to the enjoyment of a real difficulty. From the first I had observed that the Egyptian authorities did not wish to encourage English explorations of the slave-producing districts, as such examinations would be detrimental to the traffic, and would lead to reports to the European governments that would ultimately prohibit the trade. It was perfectly clear that the utmost would be done to prevent my expedition from starting. This opposition gave a piquancy to the undertaking, and I resolved that nothing should thwart my plans. Accordingly I set to work in earnest. I had taken the precaution to obtain an order upon the Treasury at Khartoum for what money I required, and as ready cash performs wonders in that country of credit and delay, I was within a few weeks ready to start. I engaged three vessels, including two large noggurs or sailing barges, and a good decked vessel with comfortable cabins, known by all Nile tourists as a diahbiah.

On December 18th, 1862, we left Khartoum. Our course up the river was slow and laborious. At times the boats had to be dragged by the men through the high reeds. It is not surprising that the ancients gave up the exploration of the Nile, when they came to the countless windings and difficulties of the marshes. The river is like an entangled skein of thread, and the voyage is tedious and melancholy beyond description. We did not reach Gondokoro until February 2d. This was merely a station of the ivory traders, occupied for two months during the year, after which time it was deserted, the boats returning to Khartoum and the expeditions again departing to the interior.

CHAPTER XIII.

Gondokoro--A mutiny quelled--Arrival of Speke and Grant--The sources of the Nile--Arab duplicity--The boy-slave's story-- Saat adopted.

Having landed all my stores, and housed my corn in some granaries belong to Koorshid Aga, I took a receipt from him for the quantity, and gave him an order to deliver one half from my depot to Speke and Grant, should they arrive at Gondokoro during my absence in the interior. I was under an apprehension that they might arrive by some route without my knowledge, while I should be penetrating south.

There were a great number of men at Gondokoro belonging to the various traders, who looked upon me with the greatest suspicion. They could not believe that simple travelling was my object, and they were shortly convinced that I was intent upon espionage in their nefarious ivory business and slave-hunting.

I had heard when at Khartoum that the most advanced trading station was fifteen days' march from Gondokoro. I now understood that the party from that station were expected to arrive at Gondokoro in a few days, and I determined to await them, as their ivory porters returning might carry my baggage and save the backs of my transport animals.

After a few days' detention at Gondokoro I saw unmistakable sign of discontent among my men, who had evidently been tampered with by the different traders' parties. One evening several of the most disaffected came to me with a complaint that they had not enough meat, and that they must be allowed to make a razzia upon the cattle of the natives to procure some oxen. This demand being of course refused, they retired, muttering in an insolent manner their determination of stealing cattle with or without my permission. I said nothing at the time, but early on the following morning I ordered the drum to beat and the men to fall in. I made them a short address, reminding them of the agreement made at Khartoum to follow me faithfully, and of the compact that had been entered into, that they were neither to indulge in slave-hunting nor in cattle-stealing. The only effect of my address was a great outbreak of insolence on the part of the ringleader of the previous evening. This fellow, named Eesur, was an Arab, and his impertinence was so violent that I immediately ordered him twenty-five lashes, as an example to the others.

Upon the vakeel's (Saati) advancing to seize him, there was a general mutiny. Many of the men threw down their guns and seized sticks, and rushed to the rescue of their tall ringleader. Saati was a little man, and was perfectly helpless. Here was an escort! These were the men upon whom I was to depend in hours of difficulty and danger on an expedition into unknown regions! These were the fellows that I had considered to be reduced "from wolves to lambs"!

I was determined not to be balked, but to insist upon the punishment of the ringleader. I accordingly went toward him with the intention of seizing him; but he, being backed by upward of forty men, had the impertinence to attack me, rushing forward with a fury that was ridiculous. To stop his blow and to knock him into the middle of the crowd was not difficult, and after a rapid repetition of the dose I disabled him, and seizing him by the throat I called to my vakeel Saati for a rope to bind him, but in an instant I had a crowd of men upon me to rescue their leader.

How the affair would have ended I cannot say; but as the scene lay within ten yards of my boat, my wife, who was ill with fever in the cabin, witnessed the whole affray, and seeing me surrounded, she rushed out, and in a few moments she was in the middle of the crowd, who at that time were endeavoring to rescue my prisoner. Her sudden appearance had a curious effect, and calling upon several of the least mutinous to assist, she very pluckily made her way up to me. Seizing the opportunity of an indecision that was for the moment evinced by the crowd, I shouted to the drummer boy to beat the drum. In an instant the drum beat, and at the top of my voice I ordered the men to "fall in." It is curious how mechanically an order is obeyed if given at the right moment, even in the midst of mutiny. Two thirds of the men fell in and formed in line, while the remainder retreated with the ringleader, Eesur, whom they led away, declaring that he was badly hurt. The affair ended in my insisting upon all forming in line, and upon the ringleader being brought forward. In this critical moment Mrs. Baker, with great tact, came forward and implored me to forgive him if he kissed my hand and begged for pardon. This compromise completely won the men, who, although a few minutes before in open mutiny, now called upon their ringleader, Eesur, to apologize and all would be right. I made them rather a bitter speech, and dismissed them.

From that moment I felt that my expedition was fated. This outbreak was an example of what was to follow. Previously to leaving Khartoum I had felt convinced that I could not succeed with such villains for escort as these Khartoumers; thus I had applied to the Egyptian authorities for a few troops, but had been refused. I was now in an awkward position. All my men had received five months' wages in advance, according to the custom of the White Nile; thus I had no control over them. There were no Egyptian authorities in Gondokoro. It was a nest of robbers, and my men had just exhibited so pleasantly their attachment to me, and their fidelity! There was no European beyond Gondokoro, thus I should be the only white man among this colony of wolves; and I had in perspective a difficult and uncertain path, where the only chance of success lay in the complete discipline of my escort and the perfect organization of the expedition. After the scene just enacted I felt sure that my escort would give me more cause for anxiety than the acknowledged hostility of the natives.

I had been waiting at Gondokoro twelve days, expecting the arrival of Debono's party from the south, with whom I wished to return. Suddenly, on the 15th of February, I heard the rattle of musketry at a great distance and a dropping fire from the south. To give an idea of the moment I must extract verbatim from my journal as written at the time.

"Guns firing in the distance; Debono's ivory porters arriving, for whom I have waited. My men rushed madly to my boat, with the report that two white men were with them who had come from the SEA! Could they be Speke and Grant? Off I ran, and soon met them in reality. Hurrah for old England! They had come from the Victoria N'yanza, from which the Nile springs . . . . The mystery of ages solved! With my pleasure of meeting them is the one disappointment, that I had not met them farther on the road in my search for them; however, the satisfaction is, that my previous arrangements had been such as would have insured my finding them had they been in a fix . . . . My projected route would have brought me vis-a-vis with them, as they had come from the lake by the course I had proposed to take . . . . All my men perfectly mad with excitement. Firing salutes as usual with ball cartridge, they shot one of my donkeys--a melancholy sacrifice as an offering at the completion of this geographical discovery."

When I first met the two explorers they were walking along the bank of the river toward my boats. At a distance of about a hundred yards I recognized my old friend Speke, and with a heart beating with joy I took off my cap and gave a welcome hurrah! as I ran toward him. For the moment he did not recognize me. Ten years' growth of beard and mustache had worked a change; and as I was totally unexpected, my sudden appearance in the centre of Africa appeared to him incredible. I hardly required an introduction to his companion, as we felt already acquainted, and after the transports of this happy meeting we walked together to my diahbiah, my men surrounding us with smoke and noise by keeping up an unremitting fire of musketry the whole way. We were shortly seated on deck under the awning, and such rough fare as could be hastily prepared was set before these two ragged, careworn specimens of African travel, whom I looked upon with feelings of pride as my own countrymen. As a good ship arrives in harbor, battered and torn by a long and stormy voyage, yet sound in her frame and seaworthy to the last, so both these gallant travellers arrived at Gondokoro. Speke appeared the more worn of the two; he was excessively lean, but in reality was in good, tough condition. He had walked the whole way from Zanzibar, never having once ridden during that wearying march. Grant was in honorable rags, his bare knees projecting through the remnants of trousers that were an exhibition of rough industry in tailor's work. He was looking tired and feverish, but both men had a fire in the eye that showed the spirit that had led them through.

They wished to leave Gondokoro as soon as possible, en route for England, but delayed their departure until the moon should be in a position for an observation for determining the longitude. My boats were fortunately engaged by me for five months, thus Speke and Grant could take charge of them to Khartoum.

At the first blush on meeting them, I had considered my expedition as terminated by having met them, and by their having accomplished the discovery of the Nile source; but upon my congratulating them with all my heart upon the honor they had so nobly earned, Speke and Grant with characteristic candor and generosity gave me a map of their route, showing that they had been unable to complete the actual exploration of the Nile, and that a most important portion still remained to be determined. It appeared that in N. lat. 2 "degrees" 17', they had crossed the Nile, which they had tracked from the Victoria Lake; but the river, which from its exit from that lake had a northern course, turned suddenly to the WEST from Karuma Falls (the point at which they crossed it at lat. 2 "degrees" 17'). They did not see the Nile again until they arrived in N. lat. 3 "degrees" 32', which was then flowing from the west-south-west. The natives and the King of Unyoro (Kamrasi) had assured them that the Nile from the Victoria N'yanza, which they had crossed at Karuma, flowed westward for several days' journey, and at length fell into a large lake called the Luta N'zige; that this lake came from the south, and that the Nile on entering the northern extremity almost immediately made its exit, and as a navigable river continued its course to the north, through the Koshi and Madi countries. Both Speke and Grant attached great importance to this lake Luta N'zige, and the former was much annoyed that it had been impossible for them to carry out the exploration. He foresaw that stay-at-home geographers, who, with a comfortable arm-chair to sit in, travel so easily with their fingers on a map, would ask him why he had not gone from such a place to such a place? why he had not followed the Nile to the Luta N'zige lake, and from the lake to Gondokoro? As it happened, it was impossible for Speke and Grant to follow the Nile from Karuma: the tribes were fighting with Kamrasi, and no strangers could have gone through the country. Accordingly they procured their information most carefully, completed their map, and laid down the reported lake in its supposed position, showing the Nile as both influent and effluent precisely as had been explained by the natives.

Speke expressed his conviction that the Luta N'zige must be a second source of the Nile, and that geographers would be dissatisfied that he had not explored it. To me this was most gratifying. I had been much disheartened at the idea that the great work was accomplished, and that nothing remained for exploration. I even said to Speke, "Does not one leaf of the laurel remain for me?" I now heard that the field was not only open, but that an additional interest was given to the exploration by the proof that the Nile flowed out of one great lake, the Victoria, but that it evidently must derive an additional supply from an unknown lake, as it entered it at the NORTHERN extremity, while the body of the lake came from the south. The fact of a great body of water such as the Luta N'zige extending in a direct line from south to north, while the general system of drainage of the Nile was from the same direction, showed most conclusively that the Luta N'zige, if it existed in the form assumed, must have an important position in the basin of the Nile.

My expedition had naturally been rather costly, and being in excellent order it would have been heartbreaking to return fruitlessly. I therefore arranged immediately for my departure, and Speke most kindly wrote in my journal such instructions as might be useful.

On the 26th of February Speke and Grant sailed from Gondokoro. Our hearts were too full to say more than a short "God bless you!" They had won their victory; my work lay all before me. I watched their boat until it turned the corner, and wished them in my heart all honor for their great achievement. I trusted to sustain the name they had won for English perseverance, and I looked forward to meeting them again in dear old England, when I should have completed the work we had so warmly planned together.

I now weighed all my baggage, and found that I had fifty-four cantars (100 lbs. each). The beads, copper, and ammunition were the terrible onus. I therefore applied to Mahommed, the vakeel of Andrea Debono, who had escorted Speke and Grant, and I begged his co-operation in the expedition. Mahommed promised to accompany me, not only to his camp at Faloro, but throughout the whole of my expedition, provided that I would assist him in procuring ivory, and that I would give him a handsome present. All was agreed upon, and my own men appeared in high spirits at the prospect of joining so large a party as that of Mahommed, which mustered about two hundred men.

At that time I really placed dependence upon the professions of Mahommed and his people; they had just brought Speke and Grant with them, and had received from them presents of a first-class double-barrelled gun and several valuable rifles. I had promised not only to assist them in their ivory expeditions, but to give them something very handsome in addition, and the fact of my having upward of forty men as escort was also an introduction, as they would be an addition to the force, which is a great advantage in hostile countries. Everything appeared to be in good trim, but I little knew the duplicity of these Arab scoundrels. At the very moment that they were most friendly, they were plotting to deceive me, and to prevent the from entering the country. They knew that, should I penetrate the interior, the IVORY TRADE of the White Nile would be no longer a mystery, and that the atrocities of the slave trade would be exposed, and most likely be terminated by the intervention of European Powers; accordingly they combined to prevent my advance, and to overthrow my expedition completely. All the men belonging to the various traders were determined that no Englishman should penetrate into the country; accordingly they fraternized with my escort, and persuaded them that I was a Christian dog that it was a disgrace for a Mahometan to serve; that they would be starved in my service, as I would not allow them to steal cattle; that they would have no slaves; and that I should lead them--God knew where--to the sea, from whence Speke and Grant had started; that they had left Zanzibar with two hundred men, and had only arrived at Gondokoro with eighteen, thus the remainder must have been killed by the natives on the road; that if they followed me and arrived at Zanzibar, I would find a ship waiting to take me to England, and I would leave them to die in a strange country. Such were the reports circulated to prevent my men from accompanying me, and it was agreed that Mahommed should fix a day for our pretended start IN COMPANY, but that he should in reality start a few days before the time appointed; and that my men should mutiny, and join his party in cattle-stealing and slave-hunting. This was the substance of the plot thus carefully concocted.

My men evinced a sullen demeanor, neglected all orders, and I plainly perceived a settled discontent upon their general expression. The donkeys and camels were allowed to stray, and were daily missing, and recovered with difficulty. The luggage was overrun with white ants, instead of being attended to every morning. The men absented themselves without leave, and were constantly in the camps of the different traders. I was fully prepared for some difficulty, but I trusted that when once on the march I should be able to get them under discipline.

Among my people were two blacks: one, "Richarn," already described as having been brought up by the Austrian Mission at Khartoum; the other, a boy of twelve years old, "Saat." As these were the only really faithful members of the expedition, it is my duty to describe them. Richarn was an habitual drunkard, but he had his good points: he was honest, and much attached to both master and mistress. He had been with me for some months, and was a fair sportsman, and being of an entirely different race from the Arabs, he kept himself apart from them, and fraternized with the boy Saat.

Saat was a boy that would do no evil. He was honest to a superlative degree, and a great exception to the natives of this wretched country. He was a native of "Fertit," and was minding his father's goats, when a child of about six years old, at the time of his capture by the Baggara Arabs. He described vividly how men on camels suddenly appeared while he was in the wilderness with his flock, and how he was forcibly seized and thrust into a large gum sack and slung upon the back of a camel. Upon screaming for help, the sack was opened, and an Arab threatened him with a knife should he make the slightest noise. Thus quieted, he was carried hundreds of miles through Kordofan to Dongola on the Nile, at which place he was sold to slave-dealers and taken to Cairo to be sold to the Egyptian government as a drummer-boy. Being too young he was rejected, and while in the dealer's hands he heard from another slave, of the Austrian Mission at Cairo, that would protect him could he only reach their asylum. With extraordinary energy for a child of six years, he escaped from his master and made his way to the Mission, where he was well received, and to a certain extent disciplined and taught as much of the Christian religion as he could understand. In company with a branch establishment of the Mission, he was subsequently located at Khartoum, and from thence was sent up the White Nile to a Mission-station in the Shillook country. The climate of tie White Nile destroyed thirteen missionaries in the short space of six months, and the boy Saat returned with the remnant of the party to Khartoum and was readmitted into the Mission. The establishment was at that time swarming with little black boys from the various White Nile tribes, who repaid the kindness of the missionaries by stealing everything they could lay their hands upon. At length the utter worthlessness of the boys, their moral obtuseness, and the apparent impossibility of improving them determined the chief of the Mission to purge his establishment from such imps, and they were accordingly turned out. Poor little Saat, the one grain of gold amid the mire, shared the same fate.

It was about a week before our departure from Khartoum that Mrs. Baker and I were at tea in the middle of the court-yard, when a miserable boy about twelve years old came uninvited to her side, and knelt down in the dust at her feet. There was something so irresistibly supplicating in the attitude of the child that the first impulse was to give him something from the table. This was declined, and he merely begged to be allowed to live with us and to be our boy. He said that he had been turned out of the Mission, merely because the Bari boys of the establishment were thieves, and thus he suffered for their sins. I could not believe it possible that the child had been actually turned out into the streets, and believing that the fault must lie in the boy, I told him I would inquire. In the mean time he was given in charge of the cook.

It happened that on the following day I was so much occupied that I forgot to inquire at the Mission, and once more the cool hour of evening arrived, when, after the intense heat of the day, we sat at table in the open court-yard. Hardly were we seated when again the boy appeared, kneeling in the dust, with his head lowered at my wife's feet, and imploring to be allowed to follow us. It was in vain that I explained that we had a boy and did not require another; that the journey was long and difficult, and that he might perhaps die. The boy feared nothing, and craved simply that he might belong to us. He had no place of shelter, no food; had been stolen from his parents, and was a helpless outcast.

The next morning, accompanied by Mrs. Baker, I went to the Mission and heard that the boy had borne an excellent character, and that it must have been BY MISTAKE that he had been turned out with the others. This being conclusive, Saat was immediately adopted. Mrs. Baker was shortly at work making him some useful clothes, and in an incredibly short time a great change was effected. As he came from the hands of the cook, after a liberal use of soap and water, and attired in trousers, blouse, and belt, the new boy appeared in a new character.

From that time he considered himself as belonging absolutely to his mistress. He was taught by her to sew. Richarn instructed him in the mysteries of waiting at table, and washing plates, etc., while I taught him to shoot, and gave him a light double-barrelled gun. This was his greatest pride.

Not only was the boy trustworthy, but he had an extraordinary amount of moral in addition to physical courage. If any complaint were made, and Saat was called as a witness, far from the shyness too often evinced when the accuser is brought face to face with the accused, such was Saat's proudest moment; and, no matter who the man might be, the boy would challenge him, regardless of all consequences.

We were very fond of this boy; he was thoroughly good, and in that land of iniquity, thousands of miles away from all except what was evil, there was a comfort in having some one innocent and faithful in whom to trust.

CHAPTER XIV.

Startling disclosures--The last hope seems gone--The Bari chief's advice--Hoping for the best--Ho for Central Africa!

We were to start upon the following Monday. Mahommed had paid me a visit, assuring me of his devotion, and begging me to have my baggage in marching order, as he would send me fifty porters on Monday, and we would move off in company. At the very moment that he thus professed, he was coolly deceiving me. He had arranged to start without me on Saturday, while he was proposing to march together on Monday. This I did not know at the time.

One morning I had returned to the tent after having, as usual, inspected the transport animals, when I observed Mrs. Baker looking extraordinarily pale, and immediately upon my arrival she gave orders for the presence of the vakeel (headman). There was something in her manner so different from her usual calm, that I was utterly bewildered when I heard her question the vakeel, whether the men were willing to march. "Perfectly ready," was the reply. "Then order them to strike the tent and load the animals; we start this moment."

The man appeared confused, but not more so than I. Something was evidently on foot, but what I could not conjecture. The vakeel wavered, and to my astonishment I heard the accusation made against him that during the night the whole of the escort had mutinously conspired to desert me, with my arms and ammunition that were in their hands, and to fire simultaneously at me should I attempt to disarm them. At first this charge was indignantly denied, until the boy Saat manfully stepped forward and declared that the conspiracy was entered into by the whole of the escort, and that both he and Richarn, knowing that mutiny was intended, had listened purposely to the conversation during the night; at daybreak the boy reported the fact to his mistress. Mutiny, robbery, and murder were thus deliberately determined.

I immediately ordered an angarep (travelling bedstead) to be placed outside the tent under a large tree. Upon this I laid five double-barrelled guns loaded with buckshot, a revolver, and a naked sabre as sharp as a razor. A sixth rifle I kept in my hands while I sat upon the angarep, with Richarn and Saat both with double- barrelled guns behind me. Formerly I had supplied each of my men with a piece of mackintosh waterproof to be tied over the locks of their guns during the march. I now ordered the drum to be beaten, and all the men to form in line in marching order, with their locks TIED UP IN THE WATERPROOF. I requested Mrs. Baker to stand behind me and point out any man who should attempt to uncover his locks when I should give the order to lay down their arms. The act of uncovering the locks would prove his intention, in which event I intended to shoot him immediately and take my chance with the rest of the conspirators.

I had quite determined that these scoundrels should not rob me of my own arms and ammunition, if I could prevent it.

The drum beat, and the vakeel himself went into the men's quarters and endeavored to prevail upon them to answer the call. At length fifteen assembled in line; the others were nowhere to be found. The locks of the arms were secured by mackintosh as ordered. It was thus impossible for any man to fire at me until he should have released his locks.

Upon assembling in line I ordered them immediately to lay down their arms. This, with insolent looks of defiance, they refused to do. "Down with your guns thus moment," I shouted, "sons of dogs!" And at the sharp click of the locks, as I quickly cocked the rifle that I held in my hands, the cowardly mutineers widened their line and wavered. Some retreated a few paces to the rear; others sat down and laid their guns on the ground, while the remainder slowly dispersed, and sat in twos or singly, under the various trees about eighty paces distant. Taking advantage of their indecision, I immediately rose and ordered my vakeel and Richarn to disarm them as they were thus scattered. Foreseeing that the time had arrived for actual physical force, the cowards capitulated, agreeing to give up their arms and ammunition if I would give them their written discharge. I disarmed them immediately, and the vakeel having written a discharge for the fifteen men present, I wrote upon each paper the word "mutineer" above my signature. None of them being able to read, and this being written in English, they unconsciously carried the evidence of their own guilt, which I resolved to punish should I ever find them on my return to Khartoum.

Thus disarmed, they immediately joined other of the traders' parties. These fifteen men were the "Jalyns" of my party, the remainder being Dongolowas-- all Arabs of the Nile, north of Khartoum. The Dongolowas had not appeared when summoned by the drum, and my vakeel being of their nation, I impressed upon him his responsibility for the mutiny, and that he would end his days in prison at Khartoum should my expedition fail.

The boy Saat and Richarn now assured me that the men had intended to fire at me, but that they were frightened at seeing us thus prepared, but that I must not expect one man of the Dongolowas to be any more faithful than the Jalyns. I ordered the vakeel to hunt up the men and to bring me their guns, threatening that if they refused I would shoot any man that I found with one of my guns in his hands.

There was no time for mild measures. I had only Saat (a mere child) and Richarn upon whom I could depend; and I resolved with them alone to accompany Mahommed's people to the interior, and to trust to good fortune for a chance of proceeding.

I was feverish and ill with worry and anxiety, and I was lying down upon my mat when I suddenly heard guns firing in all directions, drums beating, and the customary signs of either an arrival or departure of a trading party. Presently a messenger arrived from Koorshid Aga, the Circassian, to announce the departure of Mahommed's party without me, and my vakeel appeared with a message from the same people, that if I followed on their road (my proposed route) they would fire upon me and my party, as they would allow no English spies in their country.

My last hope seemed gone. No expedition had ever been more carefully planned; everything had been well arranged to insure success. My transport animals were in good condition, their saddles and pads had been made under my own inspection, my arms, ammunition, and supplies were abundant, and I was ready to march at five minutes' notice to any part of Africa; but the expedition, so costly and so carefully organized, was completely ruined by the very people whom I had engaged to protect it. They had not only deserted, but they had conspired to murder. There was no law in these wild regions but brute force; human life was of no value; murder was a pastime, as the murderer could escape all punishment. Mr. Petherick's vakeel had just been shot dead by one of his own men, and such events were too common to create much attention. We were utterly helpless, the whole of the people against us, and openly threatening. For myself personally I had no anxiety; but the fact of Mrs. Baker's being with me was my greatest care. I dared not think of her position in the event of my death among such savages as those around her. These thoughts were shared by her; but she, knowing that I had resolved to succeed, never once hinted an advice for retreat.

Richarn was as faithful as Saat, and I accordingly confided in him my resolution to leave all my baggage in charge of a friendly chief of the Baris at Gondokoro, and to take two fast dromedaries for him and Saat, and two horses for Mrs. Baker and myself, and to make a push through the hostile tribe for three days, to arrive among friendly people at "Moir," from which place I trusted to fortune. I arranged that the dromedaries should carry a few beads, ammunition, and the astronomical instruments.

Richarn said the idea was very mad; that the natives would do nothing for beads; that he had had great experience on the White Nile when with a former master, and that the natives would do nothing without receiving cows as payment; that it was of no use to be good to them, as they had no respect for any virtue but "force;" that we should most likely be murdered; but that if I ordered him to go, he was ready to obey.

I was delighted with Richarn's rough and frank fidelity. Ordering the horses to be brought, I carefully pared their feet. Their hard flinty hoofs, that had never felt a shoe, were in excellent order for a gallop, if necessary. All being ready, I sent for the chief of Gondokoro. Meanwhile a Bari boy arrived, sent by Koorshid Aga, to act as my interpreter.

The Bari chief was, as usual, smeared all over with red ochre and fat, and had the shell of a small land tortoise suspended to his elbow as an ornament. I proposed to him my plan of riding quickly through the Bari tribe to Moir. He replied, "Impossible! If I were to beat the great nogaras (drums), and call my people together to explain who you are, they would not hurt you; but there are many petty chiefs who do not obey me, and their people would certainly attack you when crossing some swollen torrent, and what could you do with only a man and a boy?"

His reply to my question concerning the value of beads corroborated Richarn's statement: nothing could be purchased for anything but cattle. The traders had commenced the system of stealing herds of cattle from one tribe to barter with the next neighbor; thus the entire country was in anarchy and confusion, and beads were of no value. My plan for a dash through the country was impracticable.

I therefore called my vakeel, and threatened him with the gravest punishment on my return to Khartoum. I wrote to Sir R. Colquhoun, H.M. Consul-General for Egypt, which letter I sent by one of the return boats, and I explained to my vakeel that the complaint to the British authorities would end in his imprisonment, and that in case of my death through violence he would assuredly be hanged. After frightening him thoroughly, I suggested that he should induce some of the mutineers, who were Dongolowas (his own tribe), many of whom were his relatives, to accompany me, in which case I would forgive them their past misconduct.

In the course of the afternoon he returned with the news that he had arranged with seventeen of the men, but that they refused to march toward the south, and would accompany me to the east if I wished to explore that part of the country. Their plea for refusing a southern route was the hostility of the Bari tribe. They also proposed a condition, that I should "LEAVE ALL MY TRANSPORT ANIMALS AND BAGGAGE BEHIND ME." To this insane request, which completely nullified their offer to start, I only replied by vowing vengeance against the vakeel.

The time was passed by the men in vociferously quarrelling among themselves during the day and in close conference with the vakeel during the night, the substance of which was reported on the following morning by the faithful Saat. The boy recounted their plot. They agreed to march to the east, with the intention of deserting me at the station of a trader named Chenooda, seven days' march from Gondokoro, in the Latooka country, whose men were, like themselves, Dongolowas; they had conspired to mutiny at that place and to desert to the slave-hunting party with my arms and ammunition, and to shoot me should I attempt to disarm them. They also threatened to shoot my vakeel, who now, through fear of punishment at Khartoum, exerted his influence to induce them to start. Altogether it was a pleasant state of things.

I was determined at all hazards to start from Gondokoro for the interior. From long experience with natives of wild countries I did not despair of obtaining an influence over my men, however bad, could I once quit Gondokoro and lead them among the wild and generally hostile tribes of the country. They would then be separated from the contagion of the slave-hunting parties, and would feel themselves dependent upon me for guidance. Accordingly I professed to believe in their promises to accompany me to the east, although I knew of their conspiracy; and I trusted that by tact and good management I should eventually thwart all their plans, and, although forced out of my intended course, should be able to alter my route and to work round from the east to my original plan of operations south. The interpreter given by Koorshid Aga had absconded; this was a great loss, as I had no means of communication with the natives except by casually engaging a Bari in the employment of the traders, to whom I was obliged to pay exorbitantly in copper bracelets for a few minutes' conversation.

A party of Koorshid's people had just arrived with ivory from the Latooka country, bringing with them a number of that tribe as porters. They were to return shortly, but they not only refused to allow me to accompany them, but they declared their intention of forcibly repelling me, should I attempt to advance by their route. This was a good excuse for my men, who once more refused to proceed. By pressure upon the vakeel they again yielded, but on condition that I would take one of the mutineers named "Bellaal," who wished to join them, but whose offer I had refused, as he had been a notorious ringleader in every mutiny. It was a sine qua non that he was to go; and knowing the character of the man, I felt convinced that it had been arranged that he should head the mutiny conspired to be enacted upon our arrival at Chenooda's camp in the Latooka country.

The plan that I had arranged was to leave all the baggage not indispensable with Koorshid Aga at Gondokoro, who would return it to Khartoum. I intended to wait until Koorshid's party should march, when I resolved to follow them, as I did not believe they would dare to oppose me by force, their master himself being friendly. I considered their threats as mere idle boasting to frighten me from an attempt to follow them; but there was another more serious cause of danger to be apprehended.

On the route between Gondokoro and Latooka there was a powerful tribe among the mountains of Ellyria. The chief of that tribe (Legge) had formerly massacred a hundred and twenty of a trader's party. He was an ally of Koorshid's people, who declared that they would raise the tribe against me, which would end in the defeat or massacre of my party. There was a difficult pass through the mountains of Ellyria which it would be impossible to force; thus my small party of seventeen men would be helpless. It would be merely necessary for the traders to request the chief of Ellyria to attack my party to insure its destruction, as the plunder of the baggage would be an ample reward.

There was no time for deliberation. Both the present and the future looked as gloomy as could be imagined; but I had always expected extraordinary difficulties, and they were, if possible, to be surmounted. It was useless to speculate upon chances. There was no hope of success in inaction, and the only resource was to drive through all obstacles without calculating the risk.

The day arrived for the departure of Koorshid's people. They commenced firing their usual signals, the drums beat, the Turkish ensign led the way, and they marched at 2 o'clock P.M., sending a polite message "DARING" me to follow them.

I immediately ordered the tent to be struck, the luggage to be arranged, the animals to be collected, and everything to be ready for the march. Richarn and Saat were in high spirits; even my unwilling men were obliged to work, and by 7 P.M. we were all ready.

We had neither guide nor interpreter. Not one native was procurable, all being under the influence of the traders, who had determined to render our advance utterly impossible by preventing the natives from assisting us. All had been threatened, and we, perfectly helpless, commenced the desperate journey in darkness about an hour after sunset.

"Where shall we go?" said the men, just as the order was given to start. "Who can travel without a guide? No one knows the road." The moon was up, and the mountain of Belignan was distinctly visible about nine miles distant. Knowing that the route lay on the east side of that mountain, I led the way, Mrs. Baker riding by my side, and the British flag following close behind us as a guide for the caravan of heavily laden camels and donkeys. And thus we started on our march into Central Africa on the 26th of March, 1863.

CHAPTER XV.

A start made at last--A forced march--Lightening the ship-- Waiting for the caravan--Success hangs in the balance--The greatest rascal in Central Africa-- Legge demands another bottle.

The country was park-like, but much parched by the dry weather. The ground was sandy, but firm, and interspersed with numerous villages, all of which were surrounded with a strong fence of euphorbia. The country was well wooded, being free from bush or jungle, but numerous trees, all evergreens, were scattered over the landscape. No natives were to be seen but the sound of their drums and singing in chorus was heard in the far distance. Whenever it is moonlight the nights are passed in singing and dancing, beating drums, blowing horns, and the population of whole villages thus congregate together.

After a silent march of two hours we saw watchfires blazing in the distance, and upon nearer approach we perceived the trader's party bivouacked. Their custom is to march only two or three hours on the first day of departure, to allow stragglers who may have lagged behind in Gondokoro to rejoin the party before morning.

We were roughly challenged by their sentries as we passed, and were instantly told "not to remain in their neighborhood." Accordingly we passed on for about half a mile in advance, and bivouacked on some rising ground above a slight hollow in which we found water.

The following morning was clear, and the mountain of Belignan, within three or four miles, was a fine object to direct our course. I could distinctly see some enormous trees at the foot of the mountain near a village, and I hastened forward, as I hoped to procure a guide who would also act as interpreter, many of the natives in the vicinity of Gondokoro having learned a little Arabic from the traders. We cantered on ahead of the party, regardless of the assurance of our unwilling men that the natives were not to be trusted, and we soon arrived beneath the shade of a cluster of most superb trees. The village was within a quarter of a mile, situated at the very base of the abrupt mountain. The natives seeing us alone had no fear, and soon thronged around us. The chief understood a few words of Arabic, and I offered a large payment of copper bracelets and beads for a guide. After much discussion and bargaining a bad-looking fellow offered to guide us to Ellyria, but no farther. This was about twenty-eight or thirty miles distant, and it was of vital importance that we should pass through that tribe before the trader's party should raise them against us. I had great hopes of outmarching the trader's party, as they would be delayed in Belignan by ivory transactions with the chief.

At that time the Turks were engaged in business transactions with tile natives; it was therefore all important that I should start immediately, and by a forced march arrive at Ellyria and get through the pass before they should communicate with the chief. I had no doubt that by paying blackmail I should be able to clear Ellyria, provided I was in advance of the Turks; but should they outmarch me, there would be no hope; a fight and defeat would be the climax. I accordingly gave orders for an IMMEDIATE start. "Load the camels, my brothers!" I exclaimed to the sullen ruffians around me; but not a man stirred except Richarn and a fellow named Sali, who began to show signs of improvement. Seeing that the men intended to disobey, I immediately set to work myself loading the animals, requesting my men not to trouble themselves, and begging them to lie down and smoke their pipes while I did the work. A few rose from the ground ashamed and assisted to load the camels, while the others declared it an impossibility for camels to travel by the road we were about to take, as the Turks had informed them that not even the donkeys could march through the thick jungles between Belignan and Ellyria.

"All right, my brothers!" I replied; "then we'll march as far as the donkeys can go, and leave both them and the baggage on the road when they can go no farther; but I GO FORWARD."

With sullen discontent the men began to strap on their belts and cartouche boxes and prepare for the start. The animals were loaded, and we moved slowly forward at 4.30 P.M. We had just started with the Bari guide that I had engaged at Belignan, when we were suddenly joined by two of the Latookas whom I had seen when at Gondokoro and to whom I had been very civil. It appeared that these follows, who were acting as porters to the Turks, had been beaten, and had therefore absconded and joined me. This was extraordinary good fortune, as I now had guides the whole way to Latooka, about ninety miles distant. I immediately gave them each a copper bracelet and some beads, and they very good-naturedly relieved the camels of one hundred pounds of copper rings, which they carried in two baskets on their heads.

We now crossed the broad dry bed of a torrent, and the banks being steep a considerable time was occupied in assisting the loaded animals in their descent. The donkeys were easily aided, their tails being held by two men while they shuffled and slid down the sandy banks; but every camel fell, and the loads had to be carried up the opposite bank by the men, and the camels reloaded on arrival. Here again the donkeys had the advantage, as without being unloaded they were assisted up the steep ascent by two men in front pulling at their ears, while others pushed behind. Altogether the donkeys were far more suitable for the country, as they were more easily loaded. The facility of loading is all-important, and I now had an exemplification of its effect upon both animals and men. The latter began to abuse the camels and to curse the father of this and the mother of that because they had the trouble of unloading them for the descent into the river's bed, while the donkeys were blessed with the endearing name of "my brother," and alternately whacked with the stick.

For some miles we passed through a magnificent forest of large trees. The path being remarkably good, the march looked propitious. This good fortune, however, was doomed to change. We shortly entered upon thick thorny jungles. The path was so overgrown that the camels could scarcely pass under the overhanging branches, and the leather bags of provisions piled upon their backs were soon ripped by the hooked thorns of the mimosa. The salt, rice, and coffee bags all sprang leaks, and small streams of these important stores issued from the rents which the men attempted to repair by stuffing dirty rags into the holes. These thorns were shaped like fishhooks; thus it appeared that the perishable baggage must soon become an utter wreck, as the great strength and weight of the camels bore all before them, and sometimes tore the branches from the trees, the thorns becoming fixed in the leather bags. Meanwhile the donkeys walked along in comfort, being so short that they and their loads were below the branches.

My wife and I rode about a quarter of a mile at the head of the party as an advance guard, to warn the caravan of any difficulty. The very nature of the country showed that it must be full of ravines, and yet I could not help hoping against hope that we might have a clear mile of road without a break. The evening had passed, and the light faded. What had been difficult and tedious during the day now became most serious; we could not see the branches of hooked thorns that over-hung the broken path. I rode in advance, my face and arms bleeding with countless scratches, while at each rip of a thorn I gave a warning shout--"Thorn!" for those behind, and a cry of "Hole!" for any deep rut that lay in the path. It was fortunately moonlight; but the jungle was so thick that the narrow track was barely perceptible; thus both camels and donkeys ran against the trunks of trees, smashing the luggage and breaking all that could be broken. Nevertheless the case was urgent; march we must at all hazards.

My heart sank whenever we cane to a deep ravine or hor; the warning cry of "halt" told those in the rear that once more the camels must be unloaded and the same fatiguing operation must be repeated. For hours we marched; the moon was sinking; the path, already dark, grew darker; the animals, overloaded even for a good road, were tired out, and the men were disheartened, thirsty, and disgusted. Everything was tired out. I had been working like a slave to assist and to cheer the men; I was also fatigued. We had marched from 4.30 P.M--it was now 1 A.M.; we had thus been eight hours and a half struggling along the path. The moon had sunk, and the complete darkness rendered a further advance impossible; therefore, on arrival at a large plateau of rock, I ordered the animals to be unloaded and both man and beast to rest.

Every one lay down supperless to sleep. Although tired, I could not rest until I had arranged some plan for the morrow. It was evident that we could not travel over so rough a country with the animals thus overloaded; I therefore determined to leave in the jungle such articles as could be dispensed with, and to rearrange all the loads.

At 4 A.M. I awoke, and lighting a lamp I tried in vain to wake any of the men, who lay stretched upon the ground like so many corpses, sound asleep.

I threw away about 100 lbs. of salt, divided the heavy ammunition more equally among the animals, rejected a quantity of odds and ends that, although most useful, could be forsaken, and by the time the men awoke, a little before sunrise, I had completed the work. We now reloaded the animals, who showed the improvement by stepping out briskly. We marched well for three hours at a pace that bade fair to keep us well ahead of the Turks, and at length we reached the dry bed of a stream, where the Latooka guides assured us we should obtain water by digging. This proved correct; but the holes were dug deep in several places, and hours passed before we could secure a sufficient supply for all the men and animals. Ascending from this place about a mile we came to the valley of Tollogo. We passed the night in a village of the friendly natives, and were off again bright and early. On reaching the extremity of the valley we had to thread our way through the difficult pass. Had the natives been really hostile they could have exterminated us in five minutes, as it was only necessary to hurl rocks from above to insure our immediate destruction. It was in this spot that a trader's party of one hundred and twenty-six men, well armed, had been massacred to a man the year previous.

Bad as the pass was, we had hope before us, as the Latookas explained that beyond this spot there was level and unbroken ground the whole way to Latooka. Could we only clear Ellyria before the Turks, I had no fear for the present; but at the very moment when success depended upon speed we were thus baffled by the difficulties of the ground. I therefore resolved to ride on in advance of my party, leaving them to overcome the difficulties of the pass by constantly unloading the animals, while I would reconnoitre in front, as Ellyria was not far distant. My wife and I accordingly rode on, accompanied only by one of the Latookas as a guide. After turning a sharp angle of the mountain, leaving the cliff abruptly rising to the left from the narrow path, we descended a ravine worse than any place we had previously encountered, and were obliged to dismount in order to lead our horses up the steep rocks on the opposite side. On arrival at the summit a lovely view burst upon us. The valley of Ellyria was about four hundred feet below, at about a mile distant. Beautiful mountains, some two or three thousand feet high, of gray granite, walled in the narrow vale, while the landscape of forest and plain was bounded at about fifty or sixty miles' distance to the east by the blue mountains of Latooka. The mountain of Ellyria was the commencement of the fine range that continued indefinitely to the south. The whole country was a series of natural forts occupied by a large population. A glance at the scene before me was quite sufficient. To FIGHT a way through a valley a quarter of a mile wide, hemmed in by high walls of rock and bristling with lances and arrows, would be impossible with my few men, encumbered by transport animals. Should the camels arrive I could march into Ellyria in twenty minutes, make the chief a large present, and pass on without halting until I cleared the Ellyria valley. At any rate I was well before the Turks, and the forced march at night, however distressing, had been successful. The great difficulty now lay in the ravine that we had just crossed; this would assuredly delay the caravan for a considerable time.

Tying our horses to a bush, we sat upon a rock beneath the shade of a small tree within ten paces of the path, and considered the best course to pursue. I hardly liked to risk an advance into Ellyria alone before the arrival of my whole party, as we had been very rudely received by the Tollogo people on the previous evening; nevertheless I thought it might be good policy to ride unattended into Ellyria, and thus to court an introduction to the chief. However, our consultation ended in a determination to wait where we then were until the caravan should have accomplished the last difficulty by crossing the ravine, when we would all march into Ellyria in company. For a long time we sat gazing at the valley before us in which our fate lay hidden, feeling thankful that we had thus checkmated the brutal Turks. Not a sound was heard of our approaching camels; the delay was most irksome. There were many difficult places that we had passed through, and each would be a source of serious delay to the animals.

At length we heard them in the distance. We could distinctly hear the men's voices, and we rejoiced that they were approaching the last remaining obstacle; that one ravine passed through, and all before would be easy. I heard the rattling of the stones as they drew nearer, and looking toward the ravine I saw emerge from the dark foliage of the trees within fifty yards of us the hated RED FLAG AND CRESCENT LEADING THE TURK'S PARTY! We were outmarched!

One by one, with scowling looks, the insolent scoundrels filed by us within a few feet, without making the customary salaam, neither noticing us in any way, except by threatening to shoot the Latooka, our guide, who had formerly accompanied them.

Their party consisted of a hundred and forty men armed with guns, while about twice as many Latookas acted as porters, carrying beads, ammunition, and the general effects of the party. It appeared that we were hopelessly beaten.

However, I determined to advance at all hazards on the arrival of my party, and should the Turks incite the Ellyria tribe to attack us, I intended, in the event of a fight, to put the first shot through the leader. To be thus beaten at the last moment was unendurable. Boiling with indignation as the insolent wretches filed past, treating me with the contempt of a dog, I longed for the moment of action, no matter what were the odds against us. At length their leader, Ibrahim, appeared in the rear of the party. He was riding on a donkey, being the last of the line, behind the flag that closed the march.

I never saw a more atrocious countenance than that exhibited in this man. A mixed breed, between a Turk sire and all Arab mother, he had the good features and bad qualities of either race--the fine, sharp, high-arched nose and large nostril, the pointed and projecting chin, rather high cheek-bones and prominent brow, overhanging a pair of immense black eyes full of expression of all evil. As he approached he took no notice of us, but studiously looked straight before him with the most determined insolence.

The fate of the expedition was at this critical moment retrieved by Mrs. Baker. She implored me to call him, to insist upon a personal explanation, and to offer him some present in the event of establishing amicable relations. I could not condescend to address the sullen scoundrel. He was in the act of passing us, and success depended upon that instant. Mrs. Baker herself called him. For the moment he made no reply; but upon my repeating the call in a loud key he turned his donkey toward us and dismounted. I ordered him to sit down, as his men were ahead and we were alone.

The following dialogue passed between us after the usual Arab mode of greeting. I said: "Ibrahim, why should we be enemies in the midst of this hostile country? We believe in the same God; why should we quarrel in this land of heathens, who believe in no God? You have your work to perform; I have mine. You want ivory; I am a simple traveller; why should we clash? If I were offered the whole ivory of the country I would not accept a single tusk, nor interfere with you in any way. Transact your business, and don't interfere with me; the country is wide enough for us both. I have a task before me, to reach a great lake--the head of the Nile. Reach it I WILL(Inshallah). No power shall drive me back. If you are hostile I will imprison you in Khartoum; if you assist me I will reward you far beyond any reward you have ever received. Should I be killed in this country, you will be suspected. You know the result: the Government would hang you on the bare suspicion. On the contrary, if you are friendly I will use my influence in any country that I discover, that you may procure its ivory for the sake of your master, Koorshid, who was generous to Captains Speke and Grant, and kind to me. Should you be hostile, I shall hold your master responsible as your employer. Should you assist me, I will befriend you both. Choose your course frankly, like a man--friend or enemy?"

Before he had time to reply, Mrs. Baker addressed him much in the same strain, telling him that he did not know what Englishmen were; that nothing would drive them back; that the British Government watched over them wherever they might be, and that no outrage could be committed with impunity upon a British subject; that I would not deceive him in any way; that I was not a trader; and that I should be able to assist him materially by discovering new countries rich in ivory, and that he would benefit himself personally by civil conduct.

He seemed confused, and wavered. I immediately promised him a new double-barrelled gun and some gold when my party should arrive, as an earnest of the future.

He replied that he did not himself wish to be hostile, but that all the trading parties, without one exception, were against me, and that the men were convinced that I was a consul in disguise, who would report to the authorities at Khartoum all the proceedings of the traders. He continued that he believed me, but that his men would not; that all people told lies in their country, therefore no one was credited for the truth. "However," said he, "do not associate with my people, or they may insult you; but go and take possession of that large tree (pointing to one in the valley of Ellyria) for yourself and people, and I will come there and speak with you. I will now join my men, as I do not wish them to know that I have been conversing with you." He then made a salaam, mounted his donkey, and rode off.

I had won him. I knew the Arab character so thoroughly that I was convinced that the tree he had pointed out, followed by the words, "I will come there and speak to you," was to be the rendezvous for the receipt of the promised gun and money.

I did not wait for the arrival of my men, but mounting our horses, my wife and I rode down the hillside with lighter spirits than we had enjoyed for some time past. I gave her the entire credit of the "ruse." Had I been alone I should have been too proud to have sought the friendship of the sullen trader, and the moment on which success depended would leave been lost.

On arrival at the grassy plain at the foot of the mountain there was a crowd of the trader's ruffians quarrelling for the shale of a few large trees that grew on the banks of the stream. We accordingly dismounted, and turning the horses to graze we took possession of a tree at some distance, under which a number of Latookas were already sitting. Not being very particular as to our society, we sat down and waited for the arrival of our party.

The natives were entirely naked, and precisely the same as the Bari. Their chief, Legge, was among them, and received a present from Ibrahim of a long red cotton shirt, and he assumed an air of great importance. Ibrahim explained to him who I was, and he immediately came to ask for the tribute he expected to receive as "blackmail" for the right of entree into his country. Of all the villainous countenances that I have ever seen, that of Legge excelled. Ferocity, avarice, and sensuality were stamped upon his face, and I immediately requested him to sit for his portrait, and in about ten minutes I succeeded in placing within my portfolio an exact likeness of about the greatest rascal that exists in Central Africa

I had now the satisfaction of seeing my caravan slowly winding down the hillside in good order, having surmounted all their difficulties.

Upon arrival my men were perfectly astonished at seeing us so near the trader's party, and still more confounded at my sending for Ibrahim to summon him to my tree, where I presented him with some English sovereigns and a double-barrelled gun. Nothing escapes the inquisitiveness of these Arabs; and the men of both parties quickly perceived that I had established an alliance in some unaccountable manner with Ibrahim. I saw the gun lately presented to him being handed from one to the other for examination, and both my vakeel and men appeared utterly confused at the sudden change.

The chief of Ellyria now came to inspect my luggage, and demanded fifteen heavy copper bracelets and a large quantity of beads. The bracelets most in demand are simple rings of copper five-eighths of an inch thick and weighing about a pound, smaller ones not being so much valued. I gave him fifteen such rings, and about ten pounds of beads in varieties, the red coral porcelain (dimiriaf) being the most acceptable. Legge was by no means satisfied; he said his belly was very big and it must be filled, which signified that his desire was great and must be gratified. I accordingly gave him a few extra copper rings; but suddenly he smelt spirits, one of the few bottles that I possessed of spirits of wine having broken in the medicine chest. Ibrahim begged me to give him a bottle to put him in a good humor, as he enjoyed nothing so much as araki. I accordingly gave him a pint bottle of the strongest spirits of wine.

To my amazement he broke off the neck, and holding his head well back he deliberately allowed the whole of the contents to trickle down his throat as innocently as though it had been simple water. He was thoroughly accustomed to it, as the traders were in the habit of bringing him presents of araki every season. He declared this to be excellent, and demanded another bottle. At that moment a violent storm of thunder and rain burst upon us with a fury well known in the tropics. The rain fell like a waterspout, and the throng immediately fled for shelter. So violent was the storm that not a man was to be seen; some sheltered themselves under the neighboring rocks, while others ran to their villages that were close by. The trader's people commenced a fusillade, firing off all their guns lest they should get wet and miss fire.

CHAPTER, XVI.

The greeting of the slave--traders--Collapse of the mutiny--African funerals--Visit from the Latooka chief--Bokke makes a suggestion--Slaughter of the Turks--Success as a prophet--Commoro's philosophy.

Although Ellyria was a rich and powerful country, we were not able to procure any provisions. The natives refused to sell, and their general behavior assured me of their capability of any atrocity had they been prompted to attack us by the Turks. Fortunately we had a good supply of meal that had been prepared for the journey prior to our departure from Gondokoro; thus we could not starve. I also had a sack of corn for the animals, a necessary precaution, as at this season there was not a blade of grass, all in the vicinity of the route having been burned.

We started on the 30th of March, at 7.30 A.M., and entered from the valley of Ellyria upon a perfectly flat country interspersed with trees. The ground was most favorable for the animals, being perfectly flat and free from ravines. We accordingly stepped along at a brisk pace, and the intense heat of the sun throughout the hottest hours of the day made the journey fatiguing for all but the camels. The latter were excellent of their class, and now far excelled the other transport animals, marching along with ease under loads of about 600 pounds each.

My caravan was at the rear of the trader's party; but the ground being good we left our people and cantered on to the advanced flag. It was curious to witness the motley assemblage in single file extending over about half a mile of ground. Several of the people were mounted on donkeys, some on oxen; the most were on foot, including all the women to the number of about sixty, who were the slaves of the trader's people. These carried heavy loads, and many, in addition to the burdens, carried children strapped to their backs in leather slings. After four or five hours' march during the intense heat, many of the overloaded women showed symptoms of distress and became footsore. The grass having been recently burned had left the sharp charred stumps, which were very trying to those whose sandals were not in the best condition.

The women were forced along by their brutal owners with sharp blows of the coorbatch, and one who was far advanced in pregnancy could at length go no further. Upon this the savage to whom she belonged belabored her with a large stick, and not succeeding in driving her before him, he knocked her down and jumped upon her. The woman's feet were swollen and bleeding, but later in the day I again saw her hobbling along in the rear by the aid of a bamboo.

After a few days' march we reached Latome, a large Latooka town, and upon our near approach we discovered crowds collected under two enormous trees. Presently guns fired, drums beat, and we perceived the Turkish flags leading a crowd of about a hundred men, who approached us with the usual salutes, every man firing off ball cartridge as fast as he could reload. My men were soon with this lot of ragamuffins, and this was the ivory or slave-trading party that they had conspired to join. They were marching toward me to honor me with a salute, which, upon close approach, ended by their holding their guns muzzle downward, and firing them almost into my feet. I at once saw through their object in giving me this reception. They had already heard from the other party exaggerated accounts of presents that their leader had received, and they were jealous at the fact of my having established confidence with a party opposed to them. The vakeel of Chenooda was the man who had from the first instigated my men to revolt and to join his party, and he at that moment had two of my deserters with him that had mutinied and joined him at Gondokoro. It had been agreed that the remainder of my men were to mutiny at this spot and to join him with MY ARMS AND AMMUNITION. This was to be the stage for the outbreak. The apparent welcome was only to throw me off my guard.

I was coldly polite, and begging them not to waste their powder, I went to the large tree that threw a beautiful shade, and we sat down, surrounded by a crowd of both natives and trader's people. Mahommed Her sent me immediately a fat ox for my people. Not to be under any obligation, I immediately gave him a double-barrelled gun. Ibrahim and his men occupied the shade of another enormous tree at about one hundred and fifty yards' distance.

The evening arrived, and my vakeel, with his usual cunning, came to ask me whether I intended to start tomorrow. He said there was excellent shooting in this neighborhood, and that Ibrahim's camp not being more than five hours' march beyond, I could at any time join him, should I think proper. Many of my men were sullenly listening to my reply, which was that we should start in company with Ibrahim. The men immediately turned their backs and swaggered insolently to the town, muttering something that I could not distinctly understand. I gave orders directly that no man should sleep in the town, but that all should be at their posts by the luggage under the tree that I occupied. At night several men were absent, and were with difficulty brought from the town by the vakeel. The whole of the night was passed by the rival parties quarrelling and fighting. At 5.30 on the following morning the drum of Ibrahim's party beat the call, and his men with great alacrity got their porters together and prepared to march. My vakeel was not to be found; my men were lying idly in the positions where they had slept, and not a man obeyed when I gave the order to prepare to start- except Richarn and Sali. I saw that the moment had arrived. Again I gave the order to the men to get up and load the animals. Not a man would move except three or four, who slowly rose from the ground and stood resting on their guns. In the mean time Richarn and Sali were bringing the camels and making them kneel by the luggage. The boy Saat was evidently expecting a row, and although engaged with the black women in packing, he kept his eyes constantly on me.

I now observed that Bellaal was standing very near me on my right, in advance of the men who had risen from the ground, and employed himself in eying me from head to foot with the most determined insolence. The fellow had his gun in his hand, and he was telegraphing by looks with those who were standing near him, while not one of the others rose from the ground, although close to me. Pretending not to notice Bellaal, who was now, as I had expected, once more the ringleader, for the third time I ordered the men to rise immediately and to load the camels. Not a man moved; but the fellow Bellaal marched up to me, and looking me straight in the face dashed the butt-end of his gun in defiance on the ground and led the mutiny. "Not a man shall go with you! Go where you like with Ibrahim, but we won't follow you nor move a step farther. The men shall not load the camels; you may employ the 'niggers' to do it, but not us."

I looked at this mutinous rascal for a moment. This was the outburst of the conspiracy, and the threats and insolence that I had been forced to pass over for the sake of the expedition all rushed before me. "Lay down your gun!" I thundered, "and load the camels!" "I won't," was his reply. "Then stop here!" I answered, at the same time lashing out as quick as lightning with my right hand upon his jaw.

He rolled over in a heap, his gun flying some yards from his hand, and the late ringleader lay apparently insensible among the luggage, while several of his friends ran to him and played the part of the Good Samaritan. Following up on the moment the advantage I had gained by establishing a panic, I seized my rifle and rushed into the midst of the wavering men, catching first one by the throat and then another, and dragging them to the camels, which I insisted upon their immediately loading. All except three, who attended to the ruined ringleader, mechanically obeyed. Richarn and Sali both shouted to them to "hurry"; and the vakeel arriving at this moment and seeing how matters stood, himself assisted, and urged the men to obey.

Ibrahim's party had started. The animals were soon loaded, and leaving the vakeel to take them in charge, we cantered on to overtake Ibrahim, having crushed the mutiny and given such an example that, in the event of future conspiracies, my men would find it difficult to obtain a ringleader. So ended the famous conspiracy that had been reported to me by both Saat and Richarn before we left Gondokoro; and so much for the threat of firing simultaneously at me and deserting my wife in the jungle. In those savage countries success frequently depends upon one particular moment; you may lose or win according to your action at that critical instant. We congratulated ourselves upon the termination of this affair, which I trusted would be the last of the mutinies.

Upon our arrival at a large town called Kattaga, my vakeel reported the desertion of five of my men to Mahommed Her's party, with their guns and ammunition. I abused both the vakeel and the men most thoroughly, and declared, "As for the mutineers who have joined the slave- hunters, Inshallah, the vultures shall pick their bones!"

This charitable wish--which, I believe, I expressed with intense hatred - was never forgotten either by my own men or by the Turks. Believing firmly in the evil eye, their superstitious fears were immediately excited.

I had noticed during the march from Latome that the vicinity of every town was announced by heaps of human remains. Bones and skulls formed a Golgotha within a quarter of a mile of every village. Some of these were in earthenware pots, generally broken; others lay strewn here and there, while a heap in the centre showed that some form had originally been observed in their disposition. This was explained by an extraordinary custom, most rigidly observed by the Latookas. Should a man be killed in battle the body is allowed to remain where it fell, and is devoured by the vultures and hyenas; but should he die a natural death he is buried in a shallow grave within a few feet of his own door, in the little courtyard that surrounds each dwelling. Funeral dances are then kept up in memory of the dead for several weeks, at the expiration of which time the body, being sufficiently decomposed, is exhumed.

The bones are cleaned and are deposited in an earthenware jar, and carried to a spot near the town which is regarded as the cemetery.

There is little difficulty in describing the toilette of the native, that of the men being limited to the one covering of the head, the body being entirely nude. It is curious to observe among these wild savages the consummate vanity displayed in their head-dresses. Every tribe has a distinct and unchanging fashion for dressing the hair, and so elaborate is the coiffure that hair-dressing is reduced to a science. European ladies would be startled at the fact that to perfect the coiffure of a man requires a period of from eight to ten years! However tedious the operation, the result is extraordinary. The Latookas wear most exquisite helmets, all of which are formed of their own hair, and are, of course, fixtures. At first sight it appears incredible; but a minute examination shows the wonderful perseverance of years in producing what must be highly inconvenient. The thick, crisp wool is woven with fine twine, formed from the bark of a tree, until it presents a thick network of felt. As the hair grows through this matted substance it is subjected to the same process, until, in the course of years, a compact substance is formed like a strong felt, about an inch and a half thick, that has been trained into the shape of a helmet. A strong rim about two inches deep is formed by sewing it together with thread, and the front part of the helmet is protected by a piece of polished copper, while a piece of the same metal, shaped like the half of a bishop's mitre and about a foot in length, forms the crest. The framework of the helmet being at length completed, it must be perfected by an arrangement of beads, should the owner of the bead be sufficiently rich to indulge in the coveted distinction. The beads most in fashion are the red and the blue porcelain, about the size of small peas. These are sewn on the surface of the felt, and so beautifully arranged in sections of blue and red that the entire helmet appears to be formed of beads; and the handsome crest of polished copper surmounted by ostrich plumes gives a most dignified and martial appearance to this elaborate head-dress. No helmet is supposed to be complete without a row of cowrie-shells stitched around the rim so as to form a solid edge.

Although the men devote so much attention to their head-dress, the woman's is extremely simple. It is a curious fact that while the men are remarkably handsome the women are exceedingly plain. They are immense creatures, few being under five feet seven in height, with prodigious limbs. They wear exceedingly long tails, precisely like those of horses, but made of fine twine and rubbed with red ochre and grease. These are very convenient when they creep into their huts on hands and knees! In addition to the tails, they wear a large flap of tanned leather in front. Should I ever visit that country again, I should take a great number of Freemasons' aprons for the women; these would be highly prized, and would create a perfect furore.

The day after my arrival in Latooka I was accommodated by the chief with a hut in a neat courtyard, beautifully clean and cemented with clay, ashes, and cow- dung. Not patronizing the architectural advantages of a doorway two feet high, I pitched my large tent in the yard and stowed all my baggage in the hut. All being arranged, I had a large Persian carpet spread upon the ground, and received the chief of Latooka in state. He was introduced by Ibrahim, and I had the advantage of his interpreter. I commenced the conversation by ordering a present to be laid on the carpet of several necklaces of valuable beads, copper bars, and colored cotton handkerchiefs. It was most amusing to witness his delight at a string of fifty little "berrets" (opal beads the size of marbles) which I had brought into the country for the first time, and which were accordingly extremely valuable. No sooner had he surveyed them with undisguised delight than he requested me to give him another string of opals for his wife, or she would be in a bad humor; accordingly a present for the lady was added to the already large pile of beads that lay heaped upon the carpet before him. After surveying his treasures with pride, he heaved a deep sigh, and turning to the interpreter he said, "What a row there will be in the family when my other wives see Bokke (his head wife) dressed up with this finery. Tell the 'Mattat' that unless he gives necklaces for each of my other wives they will fight!" Accordingly I asked him the number of ladies that made him anxious. He deliberately began to count upon his fingers, and having exhausted the digits of one hand I compromised immediately, begging him not to go through the whole of his establishment, and presented him with about three pounds of various beads to be divided among them. He appeared highly delighted, and declared his intention of sending all his wives to pay Mrs. Baker a visit. This would be an awful visitation, as each wife would expect a present for herself, and would assuredly leave either a child or a friend for whom she would beg an addition. I therefore told him that the heat was so great that we could not bear too many in the tent, but that if *Bokke*, his favorite, would appear, we should be glad to see her. Accordingly he departed, and shortly we were honored by a visit.

Bokke and her daughter were announced, and a pair of prettier savages I never saw. They were very clean; their hair was worn short, like that of all the women of the country, and plastered with red ochre and fat so as to look like vermilion; their faces were slightly tattooed on the cheeks and temples, and they sat down on the many-colored carpet with great surprise, and stared at the first white man and woman they had ever seen. We gave them both a number of necklaces of red and blue beads, and I secured Bokke's portrait in my sketch-book, obtaining a very correct likeness. She told us that Mahommed Her's men were very bad people; that they had burned and plundered one of her villages; and that one of the Latookas who had been wounded in the fight by a bullet had just died, and they were to dance for him to-morrow; if we would like to we could attend. She asked many questions; among others, how many wives I had, and was astonished to hear that I was contented with one. This seemed to amuse her immensely, and she laughed heartily with her daughter at the idea. She said that my wife would be much improved if she would extract her four front teeth from the lower jaw and wear the red ointment on her hair, according to the fashion of the country; she also proposed that she should pierce her under lip, and wear the long pointed polished crystal, about the size of a drawing-pencil, that is the "thing" in the Latooka country. No woman among the tribe who has any pretensions to being a "swell" would be without this highly-prized ornament; and one of my thermometers having come to an end, I broke the tube into three pieces, and they were considered as presents of the highest value, to be worn through the perforated under lip. Lest the piece should slip through the hole in the lip, a kind of rivet is formed by twine bound round the inner extremity, and this, protruding into the space left by the extraction of the four front teeth of the lower jaw, entices the tongue to act upon the extremity, which gives it a wriggling motion indescribably ludicrous during conversation.

It is difficult to explain real beauty. A defect in one country is a desideratum in another. Scars upon the face are, in Europe, a blemish; but here and in the Arab countries no beauty can be perfect until the cheeks or temples have been gashed. The Arabs make three gashes upon each cheek, and rub the wounds with salt and a kind of porridge (asida) to produce proud-flesh; thus every female slave captured by the slave-hunters is marked to prove her identity and to improve her charms. Each tribe has its peculiar fashion as to the position and form of the cicatrix.

The Latookas gash the temples and cheeks of their women, but do not raise the scar above the surface, as is the custom of the Arabs.

Polygamy is, of course, the general custom, the number of a man's wives depending entirely upon his wealth, precisely as would the number of his horses in England. There is no such thing as LOVE in these countries; the feeling is not understood, nor does it exist in the shape in which we understand it. Everything is practical, without a particle of romance. Women are so far appreciated as they are valuable animals. They grind the corn, fetch the water, gather firewood, cement the floors, cook the food, and propagate the race; but they are mere servants, and as such are valuable. The price of a good-looking, strong young wife, who could carry a heavy jar of water, would be ten cows; thus a man rich in cattle would be rich in domestic bliss, as he could command a multiplicity of wives. However delightful may be a family of daughters in England, they nevertheless are costly treasures; but in Latooka and throughout savage lands they are exceedingly profitable. The simple rule of proportion will suggest that if one daughter is worth ten cows, ten daughters must be worth a hundred; therefore a large family is a source of wealth: the girls bring the cows, and the boys milk them. All being perfectly naked (I mean the girls and the boys), there is no expense, and the children act as herdsmen to the flocks as in the patriarchal times. A multiplicity of wives thus increases wealth by the increase of family. I am afraid this practical state of affairs will be a strong barrier to missionary enterprise.

A savage holds to his cows and his women, but especially to his COWS. In a razzia fight he will seldom stand for the sake of his wives, but when he does fight it is to save his cattle.

One day, soon after Bokke's visit, I heard that there had been some disaster, and that the whole of Mahommed Her's party had been massacred. On the following morning I sent ten of my men with a party of Ibrahim's to Latome to make inquiries. They returned on the following afternoon, bringing with them two wounded men. It appeared the Mahommed Her had ordered his party of 110 armed men, in addition to 300 natives, to make a razzia upon a certain village among the mountains for slaves and cattle. They had succeeded in burning a village and in capturing a great number of slaves. Having descended the pass, a native gave them the route that would lead to the capture of a large herd of cattle that they had not yet discovered. They once more ascended the mountain by a different path, and arriving at the kraal they commenced driving off the vast herd of cattle. The Latookas, who had not fought while their wives and children were being carried into slavery, now fronted bravely against the muskets to defend their herds, and charging the Turks they drove them down the pass.

It was in vain that they fought; every bullet aimed at a Latooka struck a rock, behind which the enemy was hidden. Rocks, stones, and lances were hurled at them from all sides and from above. They were forced to retreat. The retreat ended in a panic and precipitate flight. Hemmed in on all sides, amid a shower of lances and stones thrown from the mountain above, the Turks fled pell-mell down the rocky and precipitous ravines. Mistaking their route, they came to a precipice from which there was no retreat. The screaming and yelling savages closed round them. Fighting was useless; the natives, under cover of the numerous detached rocks, offered no mark for an aim, while the crowd of armed savages thrust them forward with wild yells to the very verge of the great precipice about five hundred feet below. Down they fell, hurled to utter destruction by the mass of Latookas pressing onward! A few fought to the last, but one and all were at length forced, by sheer pressure, over the edge of the cliff, and met a just reward for their atrocities.

My men looked utterly cast down, and a feeling of horror pervaded the entire party. No quarter had been given by the Latookas, and upward of two hundred natives who had joined the slave-hunters in the attack had also perished with their allies. Mahommed Her had not himself accompanied his people, both he and Bellaal, my late ringleader, having remained in camp, the latter having, fortunately for him, been disabled, and placed hors de combat by the example I had made during the mutiny.

My men were almost green with awe when I asked them solemnly, "Where are the men who deserted from me?" Without answering a word they brought two of my guns and laid them at my feet. They were covered with clotted blood mixed with sand, which had hardened like cement over the locks and various portions of the barrels. My guns were all marked. As I looked at the numbers upon the stocks, I repeated aloud the names of the owners. "Are they all dead?" I asked. "All dead," the men replied. "FOOD FOR THE VULTURES?" I asked. "None of the bodies can be recovered," faltered my vakeel. "The two guns were brought from the spot by some natives who escaped, and who saw the men fall. They are all killed." "Better for them had they remained with me and done their duty. The hand of God is heavy," I replied. My men slunk away abashed, leaving the gory witnesses of defeat and death upon the ground. I called Saat and ordered him to give the two guns to Richarn to clean.

Not only my own men but the whole of Ibrahim's party were of opinion that I had some mysterious connection with the disaster that had befallen my mutineers. All remembered the bitterness of my prophecy, "The vultures will pick their bones", and this terrible mishap having occurred so immediately afterward took a strong hold upon their superstitious minds. As I passed through the camp the men would quietly exclaim, "Wah Illahi Hawaga!" (My God, Master!) To which I simply replied, "Robine fe!" (There is a God.) From that moment I observed an extraordinary change in the manner of both my people and those of Ibrahim, all of whom now paid us the greatest respect.

One day I sent for Commoro, the Latooka chief, and through my two young interpreters I had a long conversation with him on the customs of his country. I wished if possible to fathom the origin of the extraordinary custom of exhuming the body after burial, as I imagined that in this act some idea might be traced to a belief in the resurrection.

Commoro was, like all his people, extremely tall. Upon entering my tent he took his seat upon the ground, the Latookas not using stools like the other White Nile tribes. I commenced the conversation by complimenting him on the perfection of his wives and daughters in a funeral dance which had lately been held, and on his own agility in the performance, and inquired for whom the ceremony had been performed. He replied that it was for a man who had been recently killed, but no one of great importance, the same ceremony being observed for every person without distinction.

I asked him why those slain in battle were allowed to remain unburied. He said it had always been the custom, but that he could not explain it.

"But," I replied, "why should you disturb the bones of those whom you have already buried, and expose them on the outskirts of the town?"

"It was the custom of our forefathers," he answered, "therefore we continue to observe it."

"Have you no belief in a future existence after death? Is not some idea expressed in the act of exhuming the bones after the flesh is decayed?"

Commoro (loq.).--"Existence AFTER death! How can that be? Can a dead man get out of his grave, unless we dig him out?"

"Do you think man is like a beast, that dies and is ended?"

Commoro.--"Certainly. An ox is stronger than a man, but he dies, and his bones last longer; they are bigger. A man's bones break quickly; he is weak."

"Is not a man superior in sense to an ox? Has he not a mind to direct his actions?"

Commoro--"Some men are not so clever as an ox. Men must sow corn to obtain food, but the ox and wild animals can procure it without sowing."

"Do you not know that there is a spirit within you different from flesh? Do you not dream and wander in thought to distant places in your sleep? Nevertheless your body rests in one spot. How do you account for this?"

Commoro (laughing)--"Well, how do YOU account for it? It is a thing I cannot understand; it occurs to me every night."

"The mind is independent of the body. The actual body can be fettered, but the mind is uncontrollable. The body will die and will become dust or be eaten by vultures; but the spirit will exist forever."

Commoro--"Where will the spirit live?"

"Where does fire live? Cannot you produce a fire* (* The natives always produce fire by rubbing two sticks together.) by rubbing two sticks together? Yet you SEE not the fire in the wood. Has not that fire, that lies harmless and unseen in the sticks, the power to consume the whole country? Which is the stronger, the small stick that first PRODUCES the fire, or the fire itself? So is the spirit the element within the body, as the element of fire exists in the stick, the element being superior to the substance."

Commoro--"Ha! Can you explain what we frequently see at night when lost in the wilderness? I have myself been lost, and wandering in the dark I have seen a distant fire; upon approaching the fire has vanished, and I have been unable to trace the cause, nor could I find the spot."

"Have you no idea of the existence of spirits superior to either man or beast? Have you no fear of evil except from bodily causes?"

Commoro.--"I am afraid of elephants and other animals when in the jungle at night; but of nothing else."

"Then you believe in nothing--neither in a good nor evil spirit! And you believe that when you die it will be the end of body and spirit; that you are like other animals; and that there is no distinction between man and beast; both disappear, and end at death?"

Commoro.--"Of course they do."

"Do you see no difference in good and bad actions?"

Commoro.--"Yes, there are good and bad in men and beasts."

"Do you think that a good man and a bad must share the same fate, and alike die, and end?"

Commoro.--"Yes; what else can they do? How can they help dying? Good and bad all die."

"Their bodies perish, but their spirits remain; the good in happiness, the bad in misery. If you leave no belief in a future state, WHY SHOULD A MAN BE GOOD? Why should he not be bad, if he can prosper by wickedness?"

Commoro.--"Most people are bad; if they are strong they take from the weak. The good people are all weak; they are good because they are not strong enough to be bad."

Some corn had been taken out of a sack for the horses, and a few grains lying scattered on the ground, I tried the beautiful metaphor of St. Paul as an example of a future state. Making a small hole with my finger in the ground, I placed a grain within it: "That," I said, "represents you when you die." Covering it with earth, I continued, "That grain will decay, but from it will rise the plant that will produce a reappearance of the original form."

Commoro.--"Exactly so; that I understand. But the original grain does NOT rise again; it rots like the dead man, and is ended. The fruit produced is not the same grain that we buried, but the PRODUCTION of that grain. So it is with man. I die, and decay, and am ended; but my children grow up like the fruit of the grain. Some men have no children, and some grains perish without fruit; then all are ended."

I was obliged to change the subject of conversation. In this wild naked savage there was not even a superstition upon which to found a religious feeling; there was a belief in matter, and to his understanding everything was MATERIAL. It was extraordinary to find so much clearness of perception combined with such complete obtuseness to anything ideal.

CHAPTER XVII

Disease in the camp--Forward under difficulties--Our cup of misery overflows--A rain-maker in a dilemma--Fever again--Ibrahim's quandary--Firing the prairie.

Sickness now rapidly spread among my animals. Five donkeys died within a few days, and the rest looked poor. Two of my camels died suddenly, having eaten the poison-bush. Within a few days of this disaster my good old hunter and companion of all my former sports in the Base country, Tetel, died. These terrible blows to my expedition were most satisfactory to the Latookas, who ate the donkeys and other animals the moment they died. It was a race between the natives and the vultures as to who should be first to profit by my losses.

Not only were the animals sick, but my wife was laid up with a violent attack of gastric fever, and I was also suffering from daily attacks of ague. The small-pox broke out among the Turks. Several people died, and, to make matters worse, they insisted upon inoculating themselves and all their slaves; thus the whole camp was reeking with this horrible disease.

Fortunately my camp was separate and to windward. I strictly forbade my men to inoculate themselves, and no case of the disease occurred among my people; but it spread throughout the country. Small-pox is a scourge among the tribes of Central Africa, and it occasionally sweeps through the country and decimates the population.

I had a long examination of Wani, the guide and interpreter, respecting the country of Magungo. Loggo, the Bari interpreter, always described Magungo as being on a large river, and I concluded that it must be the Asua; but upon cross-examination I found he used the word "Bahr" (in Arabic signifying river or sea) instead of "Birbe (lake). This important error being discovered gave a new feature to the geography of this part. According to his description, Magungo was situated on a lake so large that no one knew its limits. Its breadth was such that, if one journeyed two days east and the same distance west, there was no land visible on either quarter, while to the south its direction was utterly unknown. Large vessels arrived at Magungo from distant arid unknown parts, bringing cowrie-shells and beads in exchange for ivory. Upon these vessels white men had been seen. All the cowrie-shells used in Latooka and the neighboring countries were supplied by these vessels, but none had arrived for the last two years.

I concluded the lake was no other than the N'yanza, which, if the position of Mangungo were correct, extended much farther north than Speke had supposed. I determined to take the first opportunity to push for Magungo. The white men spoken of by Wani probably referred to Arabs, who, being simply brown, were called white men by the blacks. I was called a VERY WHITE MAN as a distinction; but I have frequently been obliged to take off my shirt to exhibit the difference of color between myself and men, as my face had become brown.

The Turks had set June 23d as the time for their departure from Latooka. On the day preceding my wife was dangerously ill with bilious fever, and was unable to stand, and I endeavored to persuade the trader's party to postpone their departure for a few days. They would not hear of such a proposal; they had so irritated the Latookas that they feared an attack, and their captain or vakeel, Ibrahim, had ordered them immediately to vacate the country. This was a most awkward position for me. The traders had incurred the hostility of the country, and I should bear the brunt of it should I remain behind alone. Without their presence I should be unable to procure porters, as the natives would not accompany my feeble party, especially as I could offer them no other payment than beads or copper. The rain had commenced within the last few days at Latooka, and on the route toward Obbo we should encounter continual storms. We were to march by a long and circuitous route to avoid the rocky passes that would be dangerous in the present spirit of the country, especially as the traders possessed large herds that must accompany the party. They allowed five days' march for the distance to Obbo by the intended route. This was not an alluring programme for the week's entertainment, with my wife almost in a dying state! However, I set to work and fitted an angarep with arched hoops from end to end, so as to form a frame like the cap of a wagon. This I covered with two waterproof Abyssinian tanned hides securely strapped, and lashing two long poles parallel to the sides of the angarep, I formed an excellent palanquin. In this she was assisted, and we started on June 23d.

On our arrival at Obbo both my wife and I were excessively ill with bilious fever, and neither could assist the other. The old chief of Obbo, Katchiba, hearing that we were dying, came to charm us with some magic spell. He found us lying helpless, and immediately procured a small branch of a tree, and filling his mouth with water he squirted it over the leaves and about the floor of the hut. He then waved the branch around my wife's head, also around mine, and completed the ceremony by sticking it in the thatch above the doorway. He told us we should now get better, and, perfectly satisfied, took his leave.

The hut was swarming with rats and white ants, the former racing over our bodies during the night and burrowing through the floor, filling our only room with mounds like molehills. As fast as we stopped the holes, others were made with determined perseverance. Having a supply of arsenic, I gave them an entertainment, the effect being disagreeable to all parties, as the rats died in their holes and created a horrible effluvium, while fresh hosts took the place of the departed. Now and then a snake would be seen gliding within the thatch, having taken shelter front the pouring rain.

The small-pox was raging throughout the country, and the natives were dying like flies in winter. The country was extremely unhealthy, owing to the constant rain and the rank herbage, which prevented a free circulation of air, and the extreme damp induced fevers. The temperature was 65 degrees Fahr. at night and 72 degrees during the day; dense clouds obscured the sun for many days, and the air was reeking with moisture. In the evening it was always necessary to keep a blazing fire within the hut, as the floor and walls were wet and chilly.

The wet herbage disagreed with my baggage animals.

Innumerable flies appeared, including the tsetse, and in a few weeks the donkeys had no hair left, either on their ears or legs. They drooped and died one by one. It was in vain that I erected sheds and lighted fires; nothing would protect them from the flies. The moment the fires were lit the animals would rush wildly into the smoke, from which nothing would drive them; and in the clouds of imaginary protection they would remain all day, refusing food. On the 16th of July my last horse, Mouse, died. He had a very long tail, for which I obtained A COW IN EXCHANGE. Nothing was prized so highly as horses' tails, the hairs being used for stringing beads and also for making tufts as ornaments, to be suspended from the elbows. It was highly fashionable in Obbo for the men to wear such tufts formed of the bushy ends of cows' tails.

It was also "the thing" to wear six or eight polished rings of iron, fastened so tightly round the throat as almost to choke the wearer, and somewhat resembling dog-collars.

For months we dragged on a miserable existence at Obbo, wrecked by fever. The quinine was exhausted; thus the disease worried me almost to death, returning at intervals of a few days. Fortunately my wife did not suffer so much as I did. I had nevertheless prepared for the journey south, and as travelling on foot would have been impossible in our weak state, I had purchased and trained three oxen in lieu of horses. They were named "Beef," "Steaks," and "Suet." "Beef" was a magnificent animal, but having been bitten by the flies he so lost his condition that I changed his name to "Bones." We were ready to start, and the natives reported that early in January the Asua would be fordable. I had arranged with Ibrahim that he should supply me with porters for payment in copper bracelets, and that he should accompany me with one hundred men to Kamrasi's country (Unyoro) on condition that he would restrain his people from all misdemeanors, and that they should be entirely subservient to me.

It was the month of December, and during the nine, months that I had been in correspondence with his party I had succeeded in acquiring an extraordinary influence. Although my camp was nearly three quarters of a mile from their zareeba, I had been besieged daily for many months for everything that was wanted. My camp was a kind of general store that appeared to be inexhaustible. I gave all that I had with a good grace, and thereby gained the good-will of the robbers, especially as my large medicine chest contained a supply of drugs that rendered me in their eyes a physician of the first importance. I had been very successful with my patients, and the medicines that I generally used being those which produced a very decided effect, both the Turks and natives considered them with perfect faith. There was seldom any difficulty in prognosticating the effect of tartar emetic, and this became the favorite drug that was almost daily applied for, a dose of three grains enchanting the patient, who always advertised my fame by saying "He told me I should be sick, and, by Allah! there was no mistake about it." Accordingly there was a great run upon the tartar emetic.

Many people in Debono's camp had died, including several of my deserters who had joined them. News was brought that in three separate fights with the natives my deserters had been killed on every occasion, and my men and those of Ibrahim unhesitatingly declared it was the "hand of God." None of Ibrahim's men had died since we left Latooka. One man, who had been badly wounded by a lance thrust through his abdomen, I had successfully treated; and the trading party, who would at one time gladly have exterminated me, now exclaimed, "What shall we do when the Sowar (traveller) leaves the country?" Mrs. Baker had been exceedingly kind to the women and children of both the traders and natives, and together we had created so favorable an impression that we were always referred to as umpires in every dispute. My own men, although indolent, were so completely disciplined that they would not have dared to disobey an order, and they looked back upon their former mutinous conduct with surprise at their own audacity, and declared that they feared to return to Khartoum, as they were sure that I would not forgive them.

One day, hearing a great noise of voices and blowing of horns in the direction of Katchiba's residence, I sent to inquire the cause. The old chief himself appeared very angry and excited. He said that his people were very bad, that they had been making a great noise and finding fault with him because he had not supplied them with a few showers, as they wanted to sow their crop of tullaboon. There had been no rain for about a fortnight.

Well," I replied, "you are the rain-maker; why don't you give your people rain?" "Give my people rain!" said Katchiba. "I give them rain if they don't give me goats? You don't know my people. If I am fool enough to give them rain before they give me the goats, they would let me starve! No, no! let them wait. If they don't bring me supplies of corn, goats, fowls, yams, merissa, and all that I require, not one drop of rain shall ever fall again in Obbo! Impudent brutes are my people! Do yon know, they have positively threatened to kill me unless I bring the rain?

They shan't have a drop. I will wither the crops and bring a plague upon their flocks. I'll teach these rascals to insult me!"

With all this bluster, I saw that old Katchiba was in a great dilemma, and that he would give anything for a shower, but that lie did not know how to get out of the scrape. It was a common freak of the tribes to sacrifice the rain-maker should he be unsuccessful. He suddenly altered his tone, and asked, "Have you any rain in your country?" I replied that we had, every now and then. "How do you bring it? Are you a rain-maker?" I told him that no one believed in rain- makers in our country, but that we understood how to bottle lightning (meaning electricity). "I don't keep mine in bottles, but I have a houseful of thunder and lightning," he most coolly replied; "but if you can bottle lightning, you must understand rain-making. What do you think of the weather to-day?" I immediately saw the drift of the cunning old Katchiba; he wanted professional advice. I replied that he must know all about it, as he was a regular rain- maker. "Of course I do," he answered, "but I want to know what YOU think of it." "Well," I said, "I don't think we shall have any steady rain, but I think we may have a heavy shower in about four days." I said this as I had observed fleecy clouds gathering daily in the afternoon. "Just my opinion!" said Katchiba, delighted. "In four or perhaps in five days I intend to give then one shower-- just one shower. Yes, I'll just step down to them now and tell the rascals that if they will bring me some goats by this evening and some corn to-morrow morning I will give them in four or five days just one shower." To give effect to his declaration he gave several toots upon his magic whistle. "Do you use whistles in your country?" inquired Katchiba. I only replied by giving so shrill and deafening a whistle on my fingers that Katchiba stopped his ears, and relapsing into a smile of admiration he took a glance at the sky from the doorway to see if any sudden effect had been produced. "Whistle again," he said, and once more I performed like the whistle of a locomotive. "That will do; we shall have it," said the cunning old rain-maker, and proud of having so knowingly obtained "counsel's opinion" on his case, he toddled off to his impatient subjects.

In a few days a sudden storm of rain and violent thunder added to Katchiba's renown, and after the shower horns were blowing and nogaras were beating in honor of their chief. Entre nous, my whistle was considered infallible.

A bad attack of fever laid me up until the 31st of December. On the first day of January, 1864, I was hardly able to stand, and was nearly worn out at the very time that I required my strength, as we were to start south in a few days. Although my quinine had been long since exhausted, I had reserved ten grains to enable me to start in case the fever should attack me at the time of departure. I now swallowed my last dose.

It was difficult to procure porters; therefore I left all my effects at my camp in charge of two of my men, and I determined to travel light, without the tent, and to take little beyond ammunition and cooking utensils. Ibrahim left forty-five men in his zareeba, and on the 5th of January we started.

In four days' march we reached the Asua River, and on January 13th arrived at Shooa, in latitude 3 degrees 4'.

Two days after our arrival at Shooa all of our Obbo porters absconded. They had heard that we were bound for Kamrasi's country, and having received exaggerated accounts of his power from the Shooa people, they had determined upon retreat; thus we were at once unable to proceed, unless we could procure porters from Shooa. This was exceedingly difficult, as Kamrasi was well known here, and was not loved. His country was known as "Quanda," and I at once recognized the corruption of Speke's "Uganda." The slave woman "Bacheeta," who had formerly given me in Obbo so much information concerning Kamrasi's country, was to be our interpreter; but we also had the luck to discover a lad who had formerly been employed by Mahommed in Faloro, who also spoke the language of Quanda, and had learned a little Arabic.

I now discovered that the slave woman Bacheeta had formerly been in the service of a chief named Sali, who had been killed by Kamrasi. Sali was a friend of Rionga (Kamrasi's greatest enemy), and I had been warned by Speke not to set foot upon Rionga's territory, or all travelling in Unyoro would be cut off. I plainly saw that Bacheeta was in favor of Rionga, as a friend of the murdered Sali, by whom she had had two children, and that she would most likely tamper with the guide, and that we should be led to Rionga instead of to Kamrasi. There were "wheels within wheels."

It was now reported that in the last year, immediately after the departure of Speke and Grant from Gondokoro, Debono's people had marched directly to Rionga, allied themselves to him, crossed the Nile with his people, and had attacked Kamrasi's country, killing about three hundred of his men, and capturing many slaves. I now understood why they had deceived me at Gondokoro: they had obtained information of the country from Speke's people, and had made use of it by immediately attacking Kamrasi in conjunction with Rionga.

This would be a pleasant introduction for me on entering Unyoro, as almost immediately after the departure of Speke and Grant, Kamrasi had been invaded by the very people into whose hands his messengers had delivered them, when they were guided from Unyoro to the Turks' station at Faloro. He would naturally have considered that the Turks had been sent by Speke to attack him; thus the road appeared closed to all exploration, through the atrocities of Debono's people.

Many of Ibrahim's men, at hearing this intelligence, refused to proceed to Unyoro. Fortunately for me, Ibrahim had been extremely unlucky in procuring ivory. The year had almost passed away, and he had a mere nothing with which to return to Gondokoro. I impressed upon him how enraged Koorshid would be should he return with such a trifle. Already his own men declared that he was neglecting razzias because he was to receive a present from me if we reached Unyoro. This they would report to his master (Koorshid), and it would be believed should he fail in securing ivory. I guaranteed him 100 cantars (10,000 pounds) if he would push on at all hazards with me to Kamrasi and secure me porters from Shooa. Ibrahim behaved remarkably well. For some time past I had acquired a great influence over him, and he depended so thoroughly upon my opinion that he declared himself ready to do all that I suggested. Accordingly I desired him to call his men together, and to leave in Shooa all those who were disinclined to follow us.

At once I arranged for a start, lest some fresh idea should enter the ever-suspicious brains of our followers and mar the expedition. It was difficult to procure porters, and I abandoned all that was not indispensable--our last few pounds of rice and coffee, and even the great sponging-bath, that emblem of civilization that had been clung to even when the tent had been left behind.

On the 18th of January, 1864, we left Shooa. The pure air of that country had invigorated us, and I was so improved in strength that I enjoyed the excitement of the launch into unknown lands. The Turks knew nothing of the route south, and I accordingly took the lead of the entire party. I had come to a distinct understanding with Ibrahim that Kamrasi's country should belong to ME; not an act of felony would be permitted; all were to be under my government, and I would insure him at least 100 cantars of tusks.

Eight miles of agreeable march through the usual park-like country brought us to the village of Fatiko, situated upon a splendid plateau of rock upon elevated ground with beautiful granite cliffs, bordering a level table-land of fine grass that would have formed a race-course. The high rocks were covered with natives, perched upon the outline like a flock of ravens.

We halted to rest under some fine trees growing among large isolated blocks of granite and gneiss. In a short time the natives assembled around us. They were wonderfully friendly, and insisted upon a personal introduction to both myself and Mrs. Baker. We were thus compelled to hold a levee--not the passive and cold ceremony of Europe, but a most active undertaking, as each native that was introduced performed the salaam of his country by seizing both my hands and raising my arms three times to their full stretch above my head. After about one hundred Fatikos had been thus gratified by our submission to this infliction, and our arms had been subjected to at least three hundred stretches each, I gave the order to saddle the oxen immediately, and we escaped a further proof of Fatiko affection that was already preparing, as masses of natives were streaming down the rocks hurrying to be introduced. Notwithstanding the fatigue of the ceremony, I took a great fancy to these poor people. They had prepared a quantity of merissa and a sheep for our lunch, which they begged us to remain and enjoy before we started; but the pumping action of half a village not yet gratified by a presentation was too much, and mounting our oxen with aching shoulders we bade adieu to Fatiko.

On the following day our guide lost the road; a large herd of elephants had obscured it by trampling hundreds of paths in all directions. The wind was strong from the north, and I proposed to clear the country to the south by firing the prairies. There were numerous deep swamps in the bottoms between the undulations, and upon arrival at one of these green dells we fired the grass on the opposite side. In a few minutes it roared before us, and we enjoyed the grand sight of the boundless prairies blazing like infernal regions, and rapidly clearing a path south. Flocks of buzzards and the beautiful varieties of fly- catchers thronged to the dense smoke to prey upon the innumerable insects that endeavored to escape from the approaching fire.

CHAPTER XVIII

Greeting from Kamrasi's people--Suffering for the sins of others--Alone among savages--The free-masonry of Unyoro--Pottery and civilization.

After an exceedingly fatiguing march we reached the Somerset River, or Victoria White Nile, January 22d. I went to the river to see if the other side was inhabited. There were two villages on an island, and the natives came across in a canoe, bringing the BROTHER OF RIONGA. The guide, as I had feared during the journey, had deceived us, and following the secret instructions of the slave woman Bacheeta, had brought us directly to Rionga's country.

The natives at first had taken us for Mahomet Wat-el-Mek's people; but, finding their mistake, they would give us no information. We could obtain no supplies from them; but they returned to the island and shouted out that we might go to Kamrasi if we wished, but we should receive no assistance from them.

After a most enjoyable march through the exciting scenery of the glorious river crashing over innumerable falls, and in many places ornamented with rocky islands, upon which were villages and plantain groves, we at length approached the Karuma Falls, close to the village of Atada above the ferry. The heights were crowded with natives, and a canoe was sent across to within parleying distance of our side, as the roar of the rapids prevented our voices from being heard except at a short distance. Bacheeta now explained that "SPEKE'S BROTHER had arrived from his country to pay Kamrasi a visit, and had brought him valuable presents."

"Why has he brought so many men with him?" inquired the people from the canoe.

"There are so many presents for the M'Kamma (king) that he has many men to carry them," shouted Bacheeta.

"Let us look at him!" cried the headman in the boat. Having prepared for the introduction by changing my clothes in a grove of plantains for my dressing-room, and altering my costume to a tweed suit, something similar to that worn by Speke, I climbed up a high and almost perpendicular rock that formed a natural pinnacle on the face of the cliff, and waving my cap to the crowd on the opposite side, I looked almost as imposing as Nelson in Trafalgar Square.

I instructed Bacheeta, who climbed up the giddy height after me, to shout to the people that an English lady, my wife, had also arrived, and that we wished immediately to be presented to the king and his family, as we had come to thank him for his kind treatment of Speke and Grant, who had arrived safe in their own country. Upon this being explained and repeated several times the canoe approached the shore.

I ordered all our people to retire and to conceal themselves among the plantains, that the natives might not be startled by so imposing a force, while Mrs. Baker and I advanced alone to meet Kamrasi's people, who were men of some importance. Upon landing through the high reeds, they immediately recognized the similarity of my beard and general complexion to those of Speke, and their welcome was at once displayed by the most extravagant dancing and gesticulating with lances and shields, as though intending to attack, rushing at me with the points of their lances thrust close to my face, and shouting and singing in great excitement.

I made each of them a present of a bead necklace, and explained to them my wish that there should be no delay in my presentation to Kamrasi, as Speke had complained that he had been kept waiting fifteen days before the king had condescended to see him; that if this occurred no Englishman would ever visit him, as such a reception would be considered an insult. The headman replied that he felt sure I was not an impostor; but that very shortly after the departure of Speke and Grant in the previous year a number of people had arrived in their name, introducing themselves as their greatest friends. They had been ferried across the river, and well received by Kamrasi's orders, and had been presented with ivory, slaves, and leopard-skins, as tokens of friendship; but they had departed, and suddenly returned with Rionga's people, and attacked the village in which they had been so well received; and upon the country being assembled to resist them, about three hundred of Kamrasi's men had been killed in the fight. The king had therefore given orders that upon pain of death no stranger should cross the river.

He continued, "that when he saw our people marching along the bank of the river they imagined us to be the same party that had attacked them formerly, and they were prepared to resist us, and had sent on a messenger to Kamrasi, who was three days' march from Karuma, at his capital, M'rooli; until they received a reply it would be impossible to allow us to enter the country. He promised to despatch another messenger immediately to inform the king who we were, but that we must certainly wait until his return. I explained that we had nothing to eat, and that it would be very inconvenient to remain in such a spot; that I considered the suspicion displayed was exceedingly unfair, as they must see that my wife and I were white people like Speke and Grant, whereas those who had deceived them were of a totally different race, all being either black or brown.

I told him that it did not much matter; that I had very beautiful presents intended for Kamrasi, but that another great king would be only too glad to accept them, without throwing obstacles in my way. I should accordingly return with my presents.

At the same time I ordered a handsome Persian carpet, about fifteen feet square, to be displayed as one of the presents intended for the king. The gorgeous colors, as the carpet was unfolded, produced a general exclamation. Before the effect of astonishment wore off I had a basket unpacked, and displayed upon a cloth a heap of superb necklaces, that we had prepared while at Obbo, of the choicest beads, many as large as marbles, and glittering with every color of the rainbow. The garden of jewels of Aladdin's wonderful lamp could not have produced more enticing fruit. Beads were extremely rare in Kamrasi's land; the few that existed had arrived from Zanzibar, and all that I exhibited were entirely new varieties. I explained that I had many other presents, but that it was not necessary to unpack them, as we were about to return with them to visit another king, who lived some days' journey distant. "Don't go; don't go away," said the headman and his companions. "Kamrasi will –" Here an unmistakable pantomimic action explained their meaning better than words; throwing their heads well back, they sawed across their throats with their forefingers, making horrible grimaces, indicative of the cutting of throats. I could not resist laughing at the terror that my threat of returning with the presents had created. They explained that Kamrasi would not only kill them, but would destroy the entire village of Atada should we return without visiting him; but that he would perhaps punish them in precisely the same manner should they ferry us across without special orders.

"Please yourselves," I replied; "if my party is not ferried across by the time the sun reaches that spot on the heavens (pointing to the position it would occupy at about 3 P.M.) I shall return." In a state of great excitement they promised to hold a conference on the other side, and to see what arrangements could be made. They returned to Atada, leaving the whole party, including Ibrahim, exceedingly disconcerted, having nothing to eat, an impassable river before us, and five days' march of uninhabited wilderness in our rear.

The whole day passed in shouting and gesticulating our peaceful intentions to the crowd assembled on the heights on the opposite side of the river; but the boat did not return until long after the time appointed. Even then the natives would only approach sufficiently near to be heard, but nothing would induce them to land. They explained that there was a division of opinion among the people on the other side: some were in favor of receiving us, but the greater number were of opinion that we intended hostilities; therefore we must wait until orders could be sent from the king.

To assure the people of our peaceful intentions, I begged them to take Mrs. Baker and myself alone, and to leave the armed party on this side of the river until a reply should be received from Kamrasi. At this suggestion the boat immediately returned to the other side.

The day passed away, and as the sun set we perceived the canoe again paddling across the river. This time it approached directly, and the same people landed that had received the necklaces in the morning. They said that they had held a conference with the headman, and that they had agreed to receive my wife and myself, but no other person. I replied that my servants must accompany us, as we were quite as great personages as Kamrasi, and could not possibly travel without attendants. To this they demurred; therefore I dropped the subject, and proposed to load the canoe with all the presents intended for Kamrasi. There was no objection to this, and I ordered Richarn, Saat, and Ibrahim to get into the canoe to stow away the luggage as it should be handed to them, but on no account to leave the boat. I had already prepared everything in readiness, and a bundle of rifles tied up in a large blanket and 500 rounds of ball cartridge were unconsciously received on board as PRESENTS. I had instructed Ibrahim to accompany us as my servant, as he was better than most of the men in the event of a row; and I had given orders that, in case of a preconcerted signal being given, the whole force should swim the river, supporting themselves and guns upon bundles of papyrus rush. The men thought us perfectly mad, and declared that we should be murdered immediately when on the other side; however, they prepared for crossing the river in case of treachery.

At the last moment, when the boat was about to leave the shore, two of the best men jumped in with their guns. However, the natives positively refused to start; therefore, to avoid suspicion, I ordered them to retire, but I left word that on the morrow I would send the canoe across with supplies, and that one or two men should endeavor to accompany the boat to our side on every trip.

It was quite dark when we started. The canoe was formed of a large hollow tree, capable of holding twenty people, and the natives paddled us across the rapid current just below the falls. A large fire was blazing upon the opposite shore, on a level with the river, to guide us to the landing-place. Gliding through a narrow passage in the reeds, we touched the shore and landed upon a slippery rock, close to the fire, amid a crowd of people, who immediately struck up a deafening welcome with horns and flageolets, and marched us up the steep face of the rocky cliff through a dark grove of bananas. Torches led the way, followed by a long file of spearmen; then came the noisy band and ourselves, I towing my wife up the precipitous path, while my few attendants followed behind with a number of natives who had volunteered to carry the luggage.

On arrival at the top of the cliff, we were about 180 feet above the river; and after a walk of about a quarter of a mile, we were triumphantly led into the heart of the village, and halted in a small courtyard in front of the headman's residence.

Keedja waited to receive us by a blazing fire. Not having had anything to eat, we were uncommonly hungry, and to our great delight a basketful of ripe plantains was presented to us. These were the first that I had seen for many years. A gourd bottle of plantain wine was offered and immediately emptied; it resembled extremely poor cider. We were now surrounded by a mass of natives, no longer the naked savages to whom we had been accustomed, but well-dressed men, wearing robes of bark cloth, arranged in various fashions, generally like the Arab "tope" or the Roman toga. Several of the headmen now explained to us the atrocious treachery of Debono's men, who had been welcomed as friends of Speke and Grant, but who had repaid the hospitality by plundering and massacring their hosts. I assured them that no one would be more wroth than Speke when I should make him aware of the manner in which his name had been used, and that I should make a point of reporting the circumstance to the British Government. At the same time I advised them not to trust any but white people should others arrive in my name or in the names of Speke and Grant. I upheld their character as that of Englishmen, and I begged them to state if ever they had deceived them. They replied that "there could not be better men." I answered, "You MUST trust me, as I trust entirely in you, and have placed myself in your hands; but if you have ever had cause to mistrust a white man, kill me at once!--either kill me or trust in me; but let there be no suspicions."

They seemed much pleased with the conversation, and a man stepped forward and showed me a small string of blue beads that Speke bad given him for ferrying him across the river. This little souvenir of my old friend was most interesting. After a year's wandering and many difficulties, this was the first time that I had actually come upon his track. Many people told me that they had known Speke and Grant; the former bore the name of "Mollegge" (the bearded one), while Grant had been named "Masanga" (the elephant's tusk), owing to his height. The latter had been wounded at Lucknow during the Indian mutiny, and I spoke to the people of the loss of his finger. This crowned my success, as they knew without doubt that I had seen him. It was late, therefore I begged the crowd to depart, but to send a messenger the first thing in the morning to inform Kamrasi who we were, and to beg him to permit us to visit him without loss of time.

A bundle of straw was laid on the ground for Mrs. Baker and myself, and, in lieu of other beds, the ground was our resting-place. We were bitterly cold that night, as the guns were packed up in the large blanket, and, not wishing to expose them, we were contented with a Scotch plaid each. Ibrahim, Saat, and Richarn watched by turns.

On the following morning an immense crowd of natives thronged to see us. There was a very beautiful tree about a hundred yards from the village, capable of shading upward of a thousand men, and I proposed that we should sit beneath this protection and hold a conference. The headman of the village gave us a large hut with a grand doorway about seven feet high, of which my wife took possession, while I joined the crowd at the tree. There were about six hundred men seated respectfully on the ground around me, while I sat with my back to the huge knotty trunk, with Ibrahim and Richarn at a few paces distant.

The subject of conversation was merely a repetition of that of the preceding night, with the simple addition of some questions respecting the lake. Not a man would give the slightest information; the only reply, upon my forcing the question, was the pantomime already described, passing the forefinger across the throat, and exclaiming "Kamrasi!" The entire population was tongue-locked. I tried the children to no purpose: they were all dumb. White-headed old men I questioned, as to the distance of the lake from this point. They replied, "We are children; ask the old people who know the country." Never was freemasonry more secret than in the land of Unyoro. It was useless to persevere. I therefore changed the subject by saying that our people were starving on the other side, and that provisions must be sent immediately. In all savage countries the most trifling demand requires much talking. They said that provisions were scarce, and that until Kamrasi should give the order, they could give no supplies. Understanding most thoroughly the natural instincts of the natives, I told them that I must send the canoe across to fetch three oxen that I wished to slaughter. The bait took at once, and several men ran for the canoe, and we sent one of our black women across with a message to the people that three men, with their guns and ammunition, were to accompany the canoe and guide three oxen across by swimming them with ropes tied to their horns. These were the riding oxen of some of the men that it was necessary to slaughter, to exchange the flesh for flour and other supplies.

Hardly had the few boatmen departed than some one shouted suddenly, and the entire crowd sprang to their feet and rushed toward the hut where I had left Mrs. Baker. For the moment I thought that the hut was on fire, and I joined the crowd and arrived at the doorway, where I found a tremendous press to see some extraordinary sight. Every one was squeezing for the best place, and, driving them on one side, I found the wonder that had excited their curiosity. The hut being very dark, my wife had employed her solitude during my conference with the natives, in dressing her hair at the doorway, which, being very long and blonde, was suddenly noticed by some natives; a shout was given, the rush described had taken place, and the hut was literally mobbed by the crowd of savages eager to see the extraordinary novelty. The gorilla would not make a greater stir in London streets than we appeared to create at Atada.

The oxen shortly arrived; one was immediately killed, and the flesh divided into numerous small portions arranged upon the hide. Blonde hair and white people immediately lost their attractions, and the crowd turned their attention to beef. We gave them to understand that we required flour, beans, and sweet potatoes in exchange.

The market soon went briskly, and the canoe was laden with provisions and sent across to our hungry people on the other side the river.

The difference between the Unyoro people and the tribes we had hitherto seen was most striking. On the north side of the river the natives were either stark naked or wore a mere apology for clothing in the shape of a skin slung across their shoulders. The river appeared to be the limit of utter savagedom, and the people of Unyoro considered the indecency of nakedness precisely in the same light as Europeans.

Nearly all savages have some idea of earthenware; but the scale of advancement of a country between savagedom and civilization may generally be determined by the style of its pottery. The Chinese, who were as civilized as they are at the present day at a period when the English were barbarians, were ever celebrated for the manufacture of porcelain, and the difference between savage and civilized countries is always thus exemplified; the savage makes earthenware, but the civilized make porcelain; thus the gradations from the rudest earthenware will mark the improvement in the scale of civilization. The prime utensil of the African savage is a gourd, the shell of which is the bowl presented to him by nature as the first idea from which he is to model. Nature, adapting herself to the requirements of animals and man, appears in these savage countries to yield abundantly much that savage man can want. Gourds with exceedingly strong shells not only grow wild, which if divided in halves afford bowls, but great and quaint varieties form natural bottles of all sizes, from the tiny vial to the demijohn containing five gallons.

The most savage tribes content themselves with the productions of nature, confining their manufacture to a coarse and half-baked jar for carrying water; but the semi-savage, like those of Unyoro, afford an example of the first step toward manufacturing art, by their COPYING FROM NATURE. The utter savage makes use of nature--the gourd is his utensil; and the more advanced natives of Unyoro adopt it as the model for their pottery. They make a fine quality of jet-black earthenware, producing excellent tobacco-pipes most finely worked in imitation of the small egg-shaped gourd. Of the same earthenware they make extremely pretty bowls, and also bottles copied from the varieties of the bottle gourds; thus, in this humble art, we see the first effort of the human mind in manufactures, in taking nature for a model, precisely as the beautiful Corinthian capital originated in a design from a basket of flowers.

In two days reports were brought that Kamrasi had sent a large force, including several of Speke's deserters, to inspect me and see if I was really Speke's brother. I received them standing, and after thorough inspection I was pronounced to be "Speke's own brother," and all were satisfied. However, the business was not yet over; plenty of talk, and another delay of four days was declared necessary until the king should reply to the satisfactory message about to be sent. Losing all patience, I stormed, declaring Kamrasi to be mere dust, while a white man was a king in comparison. I ordered all my luggage to be conveyed immediately to the canoe, and declared that I would return immediately to my own country; that I did not wish to see any one so utterly devoid of manners as Kamrasi, and that no other white man would ever visit his kingdom.

The effect was magical! I rose hastily to depart. The chiefs implored, declaring that Kamrasi would kill them all if I retreated, to prevent which misfortune they secretly instructed the canoe to be removed. I was in a great rage, and about 400 natives, who were present, scattered in all quarters, thinking that there would be a serious quarrel. I told the chiefs that nothing should stop me, and that I would seize the canoe by force unless my whole party should be brought over from the opposite side that instant. This was agreed upon. One of Ibrahim's men exchanged and drank blood from the arm of Speke's deserter, who was Kamrasi's representative; and peace thus firmly established, several canoes were at once employed, and sixty of our men were brought across the river before sunset. The natives had nevertheless taken the precaution to send all their women away from the village.

CHAPTER XIX.

Kamrasi's cowardice--Interview with the king--The exchange of blood--The royal beggar's last chance--An astounded sovereign.

On January 31st throngs of natives arrived to carry our luggage gratis, by the king's orders. On the following day my wife became very ill, and had to be carried on a litter during the following days. On February 4th I also fell ill upon the road, and having been held on my ox by two men for some time, I at length fell into their arms and was laid under a tree for five hours. Becoming better, I rode on for two hours.

On the route we were delayed in every possible way. I never saw such cowardice as the redoubtable Kamrasi exhibited. He left his residence and retreated to the opposite side of the river, from which point he sent us false messages to delay our advance as much as possible. He had not the courage either to repel us or to receive us. On February 9th he sent word that I was to come on ALONE. I at once turned back, stating that I no longer wished to see Kamrasi, as he must be a mere fool, and I should return to my own country. This created a great stir, and messengers were at once despatched to the king, who returned an answer that I might bring all my men, but that only five of the Turks could be allowed with Ibrahim.

After a quick march of three hours through immense woods we reached the capital--a large village of grass huts situated on a barren slope. We were ferried across a river in large canoes, capable of carrying fifty men, but formed of a single tree upward of four feet wide. Kamrasi was reported to be in his residence on the opposite side; but upon our arrival at the south bank we found ourselves thoroughly deceived. We were upon a miserable flat, level with the river, and in the wet season forming a marsh at the junction of the Kafoor River with the Somerset. The latter river bounded the flat on the east, very wide and sluggish, and much overgrown with papyrus and lotus. The river we had just crossed was the Kafoor. It was perfectly dead water and about eighty yards wide, including the beds of papyrus on either side. We were shown some filthy huts that were to form our camp. The spot was swarming with mosquitoes, and we had nothing to eat except a few fowls that I had brought with me. Kamrasi was on the OTHER SIDE OF THE RIVER; they had cunningly separated us from him, and had returned with the canoes. Thus we were prisoners upon the swamp. This was our welcome from the King of Unyoro! I now heard that Speke and Grant had been lodged in this same spot.

Ibrahim was extremely nervous, as were also my men. They declared that treachery was intended, as the boats had been withdrawn, and they proposed that we should swim the river and march back to our main party, who had been left three hours in the rear. I was ill with fever, as was also my wife, and the unwholesome air of the marsh aggravated the disease. Our luggage had been left at our last station, as this was a condition stipulated by Kamrasi; thus we had to sleep upon the damp ground of the marsh in the filthy hut, as the heavy dew at night necessitated shelter. With great difficulty I accompanied Ibrahim and a few men to the bank of the river where we had landed the day before, and, climbing upon a white ant hill to obtain a view over the high reeds, I scanned the village with a telescope. The scene was rather exciting; crowds of people were rushing about in all directions and gathering from all quarters toward the river; the slope from the river to the town M'rooli was black with natives, and I saw about a dozen large canoes preparing to transport them to our side. I returned from my elevated observatory to Ibrahim, who, on the low ground only a few yards distant, could not see the opposite side of the river owing to the high grass and reeds. Without saying more, I merely begged him to mount upon the ant hill and look toward M'rooli. Hardly had he cast a glance at the scene described, than he jumped down from his stand and cried, "They arc going to attack us!" "Let us retreat to the camp and prepare for a fight!" "Let us fire at them from here as they cross in the canoes," cried others; "the buckshot will clear them off when packed in the boats." This my panic-stricken followers would have done had I not been present.

"Fools!" I said, "do you not see that the natives have no SHIELDS with them, but merely lances? Would they commence an attack without their shields? Kamrasi is coming in state to visit us." This idea was by no means accepted by my people, and we reached our little camp, and, for the sake of precaution, stationed the men in position behind a hedge of thorns. Ibrahim had managed to bring twelve picked men instead of five as stipulated; thus we were a party of twenty-four. I was of very little use, as the fever was so strong upon me that I lay helpless on the ground.

In a short time the canoes arrived, and for about an hour they were employed in crossing and recrossing, and landing great numbers of men, until they at length advanced and took possession of some huts about 200 yards from our camp. They now hallooed that Kamrasi had arrived, and, seeing some oxen with the party, I felt sure they had no evil intentions. I ordered my men to carry me in their arms to the king, and to accompany me with the presents, as I was determined to have a personal interview, although only fit for a hospital.

Upon my approach, the crowd gave way, and I was shortly laid on a mat at the king's feet. He was a fine-looking man, but with a peculiar expression of countenance, owing to his extremely prominent eyes; he was about six feet high, beautifully clean, and was dressed in a long robe of bark cloth most gracefully folded. The nails of his hands and feet were carefully attended to, and his complexion was about as dark brown as that of an Abyssinian. He sat upon a copper stool placed upon a carpet of leopard-skins, and he was surrounded by about ten of his principal chiefs.

Our interpreter, Bacheeta, now informed him who I was, and what were my intentions. He said that he was sorry I had been so long on the road, but that he had been obliged to be cautious, having been deceived by Debono's people. I replied that I was an Englishman, a friend of Speke and Grant, that they had described the reception they had met with from him, and that I had come to thank him, and to offer him a few presents in return for his kindness, and to request him to give me a guide to the Lake Luta N'zige. He laughed at the name, and repeated it several times with his chiefs. He then said it was not LUTA, but M-WOOTAN N'zige; but that it was SIX MONTHS' journey from M'rooli, and that in my weak condition I could not possibly reach it; that I should die upon the road, and that the king of my country would perhaps imagine that I had been murdered, and might invade his territory. I replied that I was weak with the toil of years in the hot countries of Africa, but that I was in search of the great lake, and should not return until I had succeeded; that I had no king, but a powerful Queen who watched over all her subjects, and that no Englishman could be murdered with impunity; therefore he should send me to the lake without delay, and there would be the less chance of my dying in his country.

I explained that the river Nile flowed for a distance of two years' journey through wonderful countries, and reached the sea, from which many valuable articles would be sent to him in exchange for ivory, could I only discover the great lake. As a proof of this, I had brought him a few curiosities that I trusted he would accept, and I regretted that the impossibility of procuring porters had necessitated the abandonment of others that had been intended for him.

I ordered the men to unpack the Persian carpet, which was spread upon the ground before him. I then gave him an Abba (large white Cashmere mantle), a red silk netted sash, a pair of scarlet Turkish shoes, several pairs of socks, a double-barrelled gun and ammunition, and a great heap of first-class beads made up into gorgeous necklaces and girdles. He took very little notice of the presents, but requested that the gun might be fired off. This was done, to the utter confusion of the crowd, who rushed away in such haste that they tumbled over each other like so many rabbits. This delighted the king, who, although himself startled, now roared with laughter. He told me that I must be hungry and thirsty; therefore he hoped I would accept something to eat and drink. Accordingly he presented me with seventeen cows, twenty pots of sour plantain cider, and many loads of unripe plantains. I inquired whether Speke had left a medicine-chest with him. He replied that it was a very feverish country, and that he and his people had used all the medicine. Thus my last hope of quinine was cut off. I had always trusted to obtain a supply from the king, as Speke had told me that he had left a bottle with him. It was quite impossible to obtain any information from him, and I was carried back to my hut, where I found Mrs. Baker lying down with fever, and neither of us could render assistance to the other.

On the following morning the king again appeared. I was better, and had a long interview. He did not appear to heed my questions, but he at once requested that I would ally myself with him, and attack his enemy, Rionga. I told him that I could not embroil myself in such quarrels, but that I had only one object, which was the lake. I requested that he would give Ibrahim a large quantity of ivory, and that on his return from Gondokoro he would bring him most valuable articles in exchange. He said that he was not sure whether my belly was black or white; by this he intended to express evil or good intentions; but that if it were white I should, of course, have no objection to exchange blood with him, as a proof of friendship and sincerity. This was rather too strong a dose! I replied that it would be impossible, as in my country the shedding of blood was considered a proof of hostility; therefore he must accept Ibrahim as my substitute. Accordingly the arms were bared and pricked. As the blood flowed it was licked by either party, and an alliance was concluded. Ibrahim agreed to act with him against all his enemies. It was arranged that Ibrahim now belonged to Kamrasi, and that henceforth our parties should be entirely separate.

On February 21st Kamrasi was civil enough to allow us to quit the marsh. My porters had by this time all deserted, and on the following day Kamrasi promised to send us porters and to allow us to start at once. There were no preparations made, however, and after some delay we were honored by a visit from Kamrasi, who promised we should start on the following day.

He concluded, as usual, by asking for my watch and for a number of beads; the latter I gave him, together with a quantity of ammunition for his guns. He showed me a beautiful double-barrelled rifle that Speke had given him. I wished to secure this to give to Speke on my return to England, as he had told me, when at Gondokoro, how he had been obliged to part with that and many other articles sorely against his will. I therefore offered to give him three common double-barrelled guns in exchange for the rifle. This he declined, as he was quite aware of the difference in quality. He then produced a large silver chronometer that he had received from Speke. "It was DEAD," he said, "and he wished me to repair it." This I declared to be impossible. He then confessed to having explained its construction and the cause of the "ticking" to his people, by the aid of a needle, and that it had never ticked since that occasion. I regretted to see such "pearls cast before swine." Thus he had plundered Speke and Grant of all they possessed before he would allow them to proceed.

It is the rapacity of the chiefs of the various tribes that renders African exploration so difficult. Each tribe wishes to monopolize your entire stock of valuables, without which the traveller would be utterly helpless. The difficulty of procuring porters limits the amount of baggage; thus a given supply must carry you through a certain period of time. If your supply should fail, the expedition terminates with your power of giving. It is thus extremely difficult to arrange the expenditure so as to satisfy all parties and still to retain a sufficient balance. Being utterly cut off from all communication with the world, there is no possibility of receiving assistance. The traveller depends entirely upon himself, under Providence, and must adapt himself and his means to circumstances.

The day of starting at length arrived. The chief and guide appeared, and we were led to the Kafoor River, where canoes were in readiness to transport us to the south side. This was to our old quarters on the marsh. The direct course to the lake was west, and I fully expected some deception, as it was impossible to trust Kamrasi. I complained to the guide, and insisted upon his pointing out the direction of the lake, which he did, in its real position, west; but he explained that we must follow the south bank of the Kafoor River for some days, as there was an impassable morass that precluded a direct course. This did not appear satisfactory, and the whole affair looked suspicious, as we had formerly been deceived by being led across the river to the same spot, and not allowed to return. We were now led along the banks of the Kafoor for about a mile, until we arrived at a cluster of huts; here we were to wait for Kamrasi, who had promised to take leave of us. The sun was overpowering, and we dismounted from our oxen and took shelter in a blacksmith's shed. In about an hour Kamrasi arrived, attended by a considerable number of men, and took his seat in our shed. I felt convinced that his visit was simply intended to peel the last skin from the onion. I had already given him nearly all that I had, but he hoped to extract the whole before I should depart.

He almost immediately commenced the conversation by asking for a pretty yellow muslin Turkish handkerchief fringed with silver drops that Mrs. Baker wore upon her head. One of these had already been given to him, and I explained that this was the last remaining, and that she required it.... He "must" have it.... It was given. He then demanded other handkerchiefs. We had literally nothing but a few most ragged towels. He would accept no excuse, and insisted upon a portmanteau being unpacked, that he might satisfy himself by actual inspection. The luggage, all ready for the journey, had to be unstrapped and examined, and the rags were displayed in succession, but so wretched and uninviting was the exhibition of the family linen that he simply returned them, and said they did not suit him. Beads he must have, or I was "his enemy." A selection of the best opal beads was immediately given him. I rose from the stone upon which I was sitting and declared that we must start immediately. "Don't be in a hurry," he replied;

"you have plenty of time; but you have not given me that watch you promised me." ... This was my only watch that he had begged for, and had been refused, every day during my stay at M'rooli. So pertinacious a beggar I had never seen. I explained to him that without the watch my journey would be useless, but that I would give him all that I had except the watch when the exploration should be completed, as I should require nothing on my direct return to Gondokoro. At the same time I repeated to him the arrangement for the journey that he had promised, begging him not to deceive me, as my wife and I should both die if we were compelled to remain another year in this country by losing the annual boats at Gondokoro.

The understanding was this: he was to give me porters to the lake, where I was to be furnished with canoes to take me to Magungo, which was situated at the junction of the Somerset. From Magungo he told me that I should see the Nile issuing from the lake close to the spot where the Somerset entered, and that the canoes should take me down the river, and porters should carry my effects from the nearest point to Shooa, and deliver me at my old station without delay. Should he be faithful to this engagement, I trusted to procure porters from Shooa, and to reach Gondokoro in time for the annual boats. I had arranged that a boat should be sent from Khartoum to await me at Gondokoro early in this year, 1864; but I felt sure that should I be long delayed, the boat would return without me, as the people would be afraid to remain alone at Gondokoro after the other boats had quitted.

In our present weak state another year of Central Africa without quinine appeared to warrant death. It was a race against time; all was untrodden ground before us, and the distance quite uncertain. I trembled for my wife, and weighed the risk of another year in this horrible country should we lose the boats. With the self-sacrificing devotion that she had shown in every trial, she implored me not to think of any risks on her account, but to push forward and discover the lake--that she had determined not to return until she had herself reached the "M'wootan N'zige."

I now requested Kamrasi to allow us to take leave, as we had not an hour to lose. In the coolest manner he replied, "I will send you to the lake and to Shooa, as I have promised, but YOU MUST LEAVE YOUR WIFE WITH ME!"

At that moment we were surrounded by a great number of natives, and my suspicions of treachery at having been led across the Kafoor River appeared confirmed by this insolent demand. If this were to be the end of the expedition, I resolved that it should also be the end of Kamrasi, and drawing my revolver quickly, I held it within two feet of his chest, and looking at him with undisguised contempt, I told him that if I touched the trigger, not all his men could save him; and that if he dared to repeat the insult I would shoot him on the spot. At the same time I explained to him that in my country such insolence would entail bloodshed, and that I looked upon him as an ignorant ox who knew no better, and that this excuse alone could save him. My wife, naturally indignant, had risen from her seat, and maddened with the excitement of the moment she made him a little speech in Arabic (not a word of which he understood), with a countenance almost as amiable as the head of Medusa. Altogether the *mine en scene utterly astonished him. The woman Bacheeta, although savage, had appropriated the insult to her mistress, and she also fearlessly let fly at Kamrasi, translating as nearly as she could the complimentary address that "Medusa" had just delivered.

Whether this little coup be theatre had so impressed Kamrasi with British female independence that he wished to be quit of his proposed bargain, I cannot say; but with an air of complete astonishment he said, "Don't be angry! I had no intention of offending you by asking for your wife. I will give your a wife, if you want one, and I thought you might have no objection to give me yours; it is my custom to give my visitors pretty wives, and I thought you might exchange. Don't make a fuss about it; if you don't like it, there's an end of it; I will never mention it again." This very practical apology I received very sternly, and merely insisted upon starting. He seemed rather confused at having committed himself, and to make amends he called his people and ordered them to carry our loads. His men ordered a number of women, who had assembled out of curiosity, to shoulder the luggage and carry it to the next village, where they would be relieved. I assisted my wife upon her ox, and with a very cold adieu to Kamrasi I turned my back most gladly on M'rooli.

CHAPTER XX.

A satanic escort--Prostrated by sun-stroke--Days and nights of sorrow - The reward for all our labor.

The country was a vast flat of grass land interspersed with small villages and patches of sweet potatoes. These were very inferior, owing to the want of drainage. For about two miles we continued on the banks of the Kafoor River. The women who carried the luggage were straggling in disorder, and my few men were much scattered in their endeavors to collect them. We approached a considerable village; but just as we were nearing it, out rushed about six hundred men with lances and shields, screaming and yelling like so many demons. For the moment I thought it was an attack, but almost immediately I noticed that women and children were mingled with the men. My men had not taken so cool a view of the excited throng that was now approaching us at full speed, brandishing their spears, and engaging with each other in mock combat. "There's a fight! there's a fight!" my men exclaimed; "we are attacked! fire at them, Ilawaga." However, in a few seconds I persuaded them that it was a mere parade, and that there was no danger.

With a rush like a cloud of locusts the natives closed around us, dancing, gesticulating, and yelling before my ox, feigning to attack us with spears and shields, then engaging in sham fights with each other, and behaving like so many madmen. A very tall chief accompanied them; and one of their men was suddenly knocked down and attacked by the crowd with sticks and lances, and lay on the ground covered with blood. What his offence had been I did not hear. The entire crowd were most grotesquely got up, being dressed in either leopard or white monkey skins, with cows' tails strapped on behind and antelopes' horns fitted upon their heads, while their chins were ornamented with false beards made of the bushy ends of cows' tails sewed together. Altogether I never saw a more unearthly set of creatures; they were perfect illustrations of my childish ideas of devils- horns, tails, and all, excepting the hoofs. They were our escort, furnished by Kamrasi to accompany us to the lake! Fortunately for all parties, the Turks were not with us on that occasion, or the Satanic escort would certainly have been received with a volley when they so rashly advanced to compliment us by their absurd performances.

We marched till 7 P.m. over flat, uninteresting country, and then halted at a miserable village which the people had deserted, as they expected our arrival. The following morning I found much difficulty in getting our escort together, as they had been foraging throughout the neighborhood; these "devil's own" were a portion of Kamrasi's troops, who considered themselves entitled to plunder ad libitum throughout the march; however, after some delay they collected, and their tall chief approached me and begged that a gun might be fired as a curiosity. The escort had crowded around us, and as the boy Saat was close to me I ordered him to fire his gun. This was Saat's greatest delight, and bang went one barrel unexpectedly, close to the tall chief's ear. The effect was charming. The tall chief, thinking himself injured, clasped his head with both hands, and bolted through the crowd, which, struck with a sudden panic, rushed away in all directions, the "devil's own" tumbling over each other and utterly scattered by the second barrel which Saat exultingly fired in derision, as Kamrasi's warlike regiment dissolved before a sound. I felt quite sure that, in the event of a fight, one scream from the "Baby," with its charge of forty small bullets, would win the battle if well delivered into a crowd of Kamrasi's troops.

On the morning of the second day we had difficulty in collecting porters, those of the preceding day having absconded; and others were recruited from distant villages by the native escort, who enjoyed the excuse of hunting for porters, as it gave them an opportunity of foraging throughout the neighborhood. During this time we had to wait until the sun was high; we thus lost the cool hours of morning, and it increased our fatigue. Having at length started, we arrived in the afternoon at the Kafoor River, at a bend from the south where it was necessary to cross over in our westerly course. The stream was in the centre of a marsh, and although deep, it was so covered with thickly-matted water-grass and other aquatic plants, that a natural floating bridge was established by a carpet of weeds about two feet thick. Upon this waving and unsteady surface the men ran quickly across, sinking merely to the ankles, although beneath the tough vegetation there was deep water.

It was equally impossible to ride or to be carried over this treacherous surface; thus I led the way, and begged Mrs. Baker to follow me on foot as quickly as possible, precisely in my track. The river was about eighty yards wide, and I had scarcely completed a fourth of the distance and looked back to see if my wife followed close to me, when I was horrified to see her standing in one spot and sinking gradually through the weeds, while her face was distorted and perfectly purple. Almost as soon as I perceived her she fell as though shot dead. In an instant I was by her side, and with the assistance of eight or ten of my men, who were fortunately close to me, I dragged her like a corpse through the yielding vegetation; and up to our waists we scrambled across to the other side, just keeping her head above the water. To have carried her would have been impossible, as we should all have sunk together through the weeds. I laid her under a tree and bathed her head and face with water, as for the moment I thought she had fainted; but she lay perfectly insensible, as though dead, with teeth and hands firmly clinched, and her eyes open but fixed. It was a coup de soleil--a sun-stroke.

Many of the porters had gone on ahead with the baggage, and I started off a man in haste to recall an angarep upon which to carry her and also for a bag with a change of clothes, as we had dragged her through the river. It was in vain that I rubbed her heart and the black women rubbed her feet to restore animation. At length the litter came, and after changing her clothes she was carried mournfully forward as a corpse. Constantly we had to halt and support her head, as a painful rattling in the throat betokened suffocation. At length we reached a village, and halted for the night.

I laid her carefully in a miserable hut, and watched beside her. I opened her clinched teeth with a small wooden wedge and inserted a wet rag, upon which I dropped water to moisten her tongue, which was dry as fur. The unfeeling brutes that composed the native escort were yelling and dancing as though all were well, and I ordered their chief at once to return with them to Kamrasi, as I would travel with them no longer. At first they refused to return, until at length I vowed that I would fire into them should they accompany us on the following morning. Day broke, and it was a relief to have got rid of the brutal escort. They had departed, and I had now my own men and the guides supplied by Kamrasi.

There was nothing to eat in this spot. My wife had never stirred since she fell by the coup de soleil, and merely respired about five times in a minute. It was impossible to remain; the people would have starved. She was laid gently upon her litter, and we started forward on our funereal course. I was ill and broken- hearted, and I followed by her side through the long day's march over wild park lands and streams, with thick forest and deep marshy bottoms, over undulating hills and through valleys of tall papyrus rushes, which, as we brushed through them on our melancholy way, waved over the litter like the black plumes of a hearse.

We halted at a village, and again the night was passed in watching. I was wet and coated with mud from the swampy marsh, and shivered with ague; but the cold within was greater than all. No change had taken place; she had never moved. I had plenty of fat, and I made four balls of about half a pound, each of which would burn for three hours. A piece of a broken water-jar formed a lamp, several pieces of rag serving for wicks. So in solitude the still calm night passed away as I sat by her side and watched. In the drawn and distorted features that lay before me I could hardly trace the same face that for years had been my comfort through all the difficulties and dangers of my path. Was she to die? Was so terrible a sacrifice to be the result of my selfish exile?

Again the night passed away. Once more the march. Though weak and ill, and for two nights without a moment's sleep, I felt no fatigue, but mechanically followed by the side of the litter as though in a dream. The same wild country diversified with marsh and forest! Again we halted. The night came, and I sat by her side in a miserable hut, with the feeble lamp flickering while she lay as in death. She had never moved a muscle since she fell. My people slept. I was alone, and no sound broke the stillness of the night. The ears ached at the utter silence, till the sudden wild cry of a hyena made me shudder as the horrible thought rushed through my brain that, should she be buried in this lonely spot, the hyena--would disturb her rest.

The morning was not far distant; it was past four o'clock. I had passed the night in replacing wet cloths upon her head and moistening her lips, as she lay apparently lifeless on her litter. I could do nothing more; in solitude and abject misery in that dark hour, in a country of savage heathen, thousands of miles away from a Christian land, I beseeched an aid above all human, trusting alone to Him.

The morning broke; my lamp had just burned out, and cramped with the night's watching I rose from my low seat and seeing that she lay in the same unaltered state I went to the door of the hut to breathe one gasp of the fresh morning air. I was watching the first red streak that heralded the rising sun, when I was startled by the words, "Thank God," faintly uttered behind me. Suddenly she had awoke from her torpor, and with a heart overflowing I went to her bedside. Her eyes were full of madness! She spoke, but the brain was gone!

I will not inflict a description of the terrible trial of seven days of brain fever, with its attendant horrors. The rain poured in torrents, and day after day we were forced to travel for want of provisions, not being able to remain in one position. Every now and then we shot a few guinea-fowl, but rarely; there was no game, although the country was most favorable. In the forests we procured wild honey, but the deserted villages contained no supplies, as we were on the frontier of Uganda, and M'tese's people had plundered the district. For seven nights I had not slept, and although as weak as a reed, I had marched by the side of her litter. Nature could resist no longer. We reached a village one evening. She had been in violent convulsions successively; it was all but over. I laid her down on her litter within a hat, covered her with a Scotch plaid, and fell upon my mat insensible, worn out with sorrow and fatigue. My men put a new handle to the pickaxe that evening, and sought for a dry spot to dig her grave!

The sun had risen when I woke. I had slept, and horrified as the idea flashed upon me that she must be dead and that I had not been with her, I started up. She lay upon her bed, pale as marble, and with that calm serenity that the features assume when the cares of life no longer act upon the mind and the body rests in death. The dreadful thought bowed me down; but as I gazed upon her in fear her chest gently heaved, not with the convulsive throbs of fever, but naturally. She was asleep; and when at a sudden noise she opened her eyes, they were calm and clear. She was saved! When not a ray of hope remained, God alone knows what helped us. The gratitude of that moment I will not attempt to describe.

Fortunately there were many fowls in this village. We found several nests of fresh eggs in the straw which littered the hut; these were most acceptable after our hard fare, and produced a good supply of soup. Having rested for two days we again moved forward, Mrs. Baker being carried on a litter.

The next day we reached the village of Parkani. For several days past our guides had told us that we were very near to the lake, and we were now assured that we should reach it on the morrow. I had noticed a lofty range of mountains at an immense distance west, and I had imagined that the lake lay on the other side of this chain; but I was now informed that those mountains formed the western frontier of the M'wootan N'zige, and that the lake was actually within a day's march of Parkani. I could not believe it possible that we were so near the object of our search. The guide Rabonga now appeared, and declared that if we started early on the following morning we should be able to wash in the lake by noon!

That night I hardly slept. For years I had striven to reach the "sources of the Nile." In my nightly dreams during that arduous voyage I had always failed, but after so much hard work and perseverance the cup was at my very lips, and I was to DRINK at the mysterious fountain before another sun should set--at that great reservoir of nature that ever since creation had baffled all discovery.

I had hoped, and prayed, and striven through all kinds of difficulties, in sickness, starvation, and fatigue, to reach that hidden source; and when it had appeared impossible we had both determined to die upon the road rather than return defeated. Was it possible that it was so near, and that tomorrow we could say, "The work is accomplished"?

The sun had not risen when I was spurring my ox after the guide, who, having been promised a double handful of beads on arrival at the lake, had caught the enthusiasm of the moment. The day broke beautifully clear, and having crossed a deep valley between the hills, we toiled up the opposite slope. I hurried to the summit. The glory of our prize burst suddenly upon me! There, like a sea of quicksilver, lay far beneath the grand expanse of water--a boundless sea horizon on the south and south-west, glittering in the noonday sun; and in the west, at fifty or sixty miles' distance, blue mountains rose from the bosom of the lake to a height of about 7000 feet above its level.

It is impossible to describe the triumph of that moment. Here was the reward for all our labor--for the years of tenacity with which we had toiled through Africa. England had won the sources of the Nile! Long before I reached this spot I had arranged to give three cheers with all our men in English style in honor of the discovery; but now that I looked down upon the great inland sea lying nestled in the very heart of Africa, and thought how vainly mankind had sought these sources throughout so many ages, and reflected that I had been the humble instrument permitted to unravel this portion of the great mystery when so many greater than I had failed, I felt too serious to vent my feelings in vain cheers for victory, and I sincerely thanked

God for having guided and supported us through all dangers to the good end. I was about 1500 feet above the lake, and I looked down from the steep granite cliff upon those welcome waters--upon that vast reservoir which nourished Egypt and brought fertility where all was wilderness--upon that great source so long hidden from mankind, that source of bounty and of blessings to millions of human beings; and as one of the greatest objects in nature, I determined to honor it with a great name. As an imperishable memorial of one loved and mourned by our gracious Queen and deplored by every Englishman, I called this great lake "the Albert N'yanza." The Victoria and the Albert lakes are the two Sources of the Nile.

The zigzag path to descend to the lake was so steep and dangerous that we were forced to leave our oxen with a guide, who was to take them to Magungo and wait for our arrival. We commenced the descent of the steep pass on foot. I led the way, grasping a stout bamboo. My wife in extreme weakness tottered down the pass, supporting herself upon my shoulder, and stopping to rest every twenty paces. After a toilsome descent of about two hours, weak with years of fever, but for the moment strengthened by success, we gained the level plain below the cliff. A walk of about a mile through flat sandy meadows of fine turf interspersed with trees and bushes brought us to the water's edge. The waves were rolling upon a white pebbly beach; I rushed into the lake, and thirsty with heat and fatigue, with a heart full of gratitude, I drank deeply from the Sources of the Nile.

CHAPTER XXI.

The cradle of the Nile--Arrival at Magungo--The blind leading the blind--Murchison Falls.

The beach was perfectly clean sand, upon which the waves rolled like those of the sea, throwing up weeds precisely as seaweed may be seen upon the English shore. It was a grand sight to look upon this vast reservoir of the mighty Nile and to watch the heavy swell tumbling upon the beach, while far to the south-west the eye searched as vainly for a bound as though upon the Atlantic. It was with extreme emotion that I enjoyed this glorious scene. My wife, who had followed me so devotedly, stood by my side pale and exhausted--a wreck upon the shores of the great Albert Lake that we had so long striven to reach. No European foot had ever trod upon its sand, nor had the eyes of a white man ever scanned its vast expanse of water. We were the first; and this was the key to the great secret that even Julius Caesar yearned to unravel, but in vain. Here was the great basin of the Nile that received EVERY DROP OF WATER, even from the passing shower to the roaring mountain torrent that drained from Central Africa toward the north. This was the great reservoir of the Nile!

The first coup d'oeil from the summit of the cliff 1500 feet above the level had suggested what a closer examination confirmed. The lake was a vast depression far below the general level of the country, surrounded by precipitous cliffs, and bounded on the west and south-west by great ranges of mountains from five to seven thousand feet above the level of its waters--thus it was the one great reservoir into which everything MUST drain; and from this vast rocky cistern the Nile made its exit, a giant in its birth. It was a grand arrangement of nature for the birth of so mighty and important a stream as the river Nile. The Victoria N'yanza of Speke formed a reservoir at a high altitude, receiving a drainage from the west by the Kitangule River; and Speke had seen the M'fumbiro Mountain at a great distance as a peak among other mountains from which the streams descended, which by uniting formed the main river Kitangule, the principal feeder of the Victoria Lake from the west, in about 2 degrees S. latitude. Thus the same chain of mountains that fed the Victoria on the east must have a watershed to the west and north that would flow into the Albert Lake. The general drainage of the Nile basin tending from south to north, and the Albert Lake extending much farther north than the Victoria, it receives the river from the latter lake, and thus monopolizes the entire head-waters of the Nile. The Albert is the grand reservoir, while the Victoria is the eastern source. The parent streams that form these lakes are from the same origin, and the Kitangule sheds its waters to the Victoria to be received EVENTUALLY by the Albert, precisely as the highlands of M'fumbiro and the Blue Mountains pour their northern drainage DIRECTLY into the Albert Lake.

That many considerable affluents flow into the Albert Lake there is no doubt. The two waterfalls seen by telescope upon the western shore descending from the Blue Mountains must be most important streams, or they could not have been distinguished at so great a distance as fifty or sixty miles. The natives assured me that very many streams, varying in size, descended the mountains upon all sides into the general reservoir.

It was most important that we should hurry forward on our journey, as our return to England depended entirely upon the possibility of reaching Gondokoro before the end of April, otherwise the boats would have departed. I started off Rabonga, to Magungo, where he was to meet us with riding oxen.

We were encamped at a small village on the shore of the lake, called Vacovia. On the following morning not one of our party could rise from the ground. Thirteen men, the boy Saat, four women, besides my wife and me, were all down with fever. The natives assured us that all strangers suffered in a like manner. The delay in supplying boats was most annoying, as every hour was precious. The lying natives deceived us in every possible manner, delaying us purposely in hope of extorting beads.

The latitude of Vacovia was 1"degree" 15' N.; longitude 30 "degrees" 50' E. My farthest southern point on the road from M'rooli was latitude 1 "degree" 13'. We were now to turn our faces toward the north, and every day's journey would bring us nearer home. But where was home? As I looked at the map of the world, and at the little red spot that represented old England far, far away, and then gazed on the wasted form and haggard face of my wife and at my own attenuated frame, I hardly dared hope for home again. We had now been three years ever toiling onward, and having completed the exploration of all the Abyssinian affluents of the Nile, in itself an arduous undertaking, we were now actually at the Nile head. We had neither health nor supplies, and the great journey lay all before us.

Eight days were passed at Vacovia before we could obtain boats, which, when they did come, proved to be mere trees neatly hollowed out in the shape of canoes. At last we were under way, and day after day we journeyed along the shore of the lake, stopping occasionally at small villages, and being delayed now and then by deserting boatmen.

The discomforts of this lake voyage were great; in the day we were cramped in our small cabin like two tortoises in one shell, and at night it almost invariably rained. We were accustomed to the wet, but no acclimatization can render the European body mosquito-proof; thus we had little rest. It was hard work for me; but for my unfortunate wife, who had hardly recovered from her attack of coup de soleil, such hardships were most distressing.

On the thirteenth day from Vacovia we found ourselves at the end of our lake voyage. The lake at this point was between fifteen and twenty miles across, and the appearance of the country to the north was that of a delta. The shores upon either side were choked with vast banks of reeds, and as the canoe skirted the edge of that upon the east coast we could find no bottom with a bamboo of twenty-five feet in length, although the floating mass appeared like terra firma. We were in a perfect wilderness of vegetation. On the west were mountains about 4000 feet above the lake level, a continuation of the chain that formed the western shore from the south. These mountains decreased in height toward the north, in which direction the lake terminated in a broad valley of reeds.

We were informed that we had arrived at Magungo, and after skirting the floating reeds for about a mile we entered a broad channel, which we were told was the embouchure of the Somerset River from Victoria N'yanza. In a short time we landed at Magungo, where we were welcomed by the chief and by our guide Rabonga, who had been sent in advance to procure oxen.

The exit of the Nile from the lake was plain enough, and if the broad channel of dead water were indeed the entrance of the Victoria Nile (Somerset), the information obtained by Speke would be remarkably confirmed. But although the chief of Magungo and all the natives assured me that the broad channel of dead water at my feet was positively the brawling river that I had crossed below the Karuma Falls, I could not understand how so fine a body of water as that had appeared could possibly enter the Albert Lake as dead water. The guide and natives laughed at my unbelief, and declared that it was dead water for a considerable distance from the junction with the lake, but that a great waterfall rushed down from a mountain, and that beyond that fall the river was merely a succession of cataracts throughout the entire distance of about six days' march to Karuma Falls. My real wish was to descend the Nile in canoes from its exit from the lake with my own men as boatmen, and thus in a short time to reach the cataracts in the Madi country; there to forsake the canoes and all my baggage, and to march direct to Gondokoro with only our guns and ammunition. I knew from native report that the Nile was navigable as far as the Madi country to about Miani's tree, which Speke had laid down by astronomical observation in lat. 3 "degrees" 34'. This would be only seven days' march from Gondokoro, and by such a direct course I should be sure to arrive in time for the boats to Khartoum.

I had promised Speke that I would explore most thoroughly the doubtful portion of the river that he had been forced to neglect from Karuma Falls to the lake. I was myself confused at the dead-water junction; and although I knew that the natives must be right--as it was their own river, and they had no inducement to mislead me--I was determined to sacrifice every other wish in order to fulfil my promise, and thus to settle the Nile question most absolutely. That the Nile flowed out of the lake I had heard, and I had also confirmed by actual inspection; from Magungo I looked upon the two countries, Koshi and Madi, through which it flowed, and these countries I must actually pass through and again meet the Nile before I could reach Gondokoro. Thus the only point necessary to settle was the river between the lake and the Karuma Falls.

The boats being ready, we took leave of the chief of Magungo, leaving him an acceptable present of beads, and descended the hill to the river, thankful at having so far successfully terminated the expedition as to have traced the lake to that important point, Magungo, which had been our clew to the discovery even so far away in time and place as the distant country of Latooka. We were both very weak and ill, and my knees trembled beneath me as we walked down the easy descent. I, in my enervated state, endeavoring to assist my wife, we were the "blind leading the blind;" but had life closed on that day we could have died most happily, for the hard fight through sickness and misery had ended in victory; and although I looked to home as a paradise never to be regained, I could have lain down to sleep in contentment on this spot, with the consolation that, if the body had been vanquished, we died with the prize in our grasp.

On arrival at the canoes we found everything in readiness, and the boatmen already in their places. Once in the broad channel of dead water we steered due east, and made rapid way until the evening. The river as it now appeared, although devoid of current, was on an average about 500 yards in width. Before we halted for the night I was subjected to a most severe attack of fever, and upon the boat reaching a certain spot I was carried on a litter, perfectly unconscious, to a village, attended carefully by my poor sick wife, who, herself half dead, followed me on foot through the marches in pitch darkness, and watched over me until the morning. At daybreak I was too weak to stand, and we were both carried down to the canoes, and crawling helplessly within our grass awning we lay down like logs while the canoes continued their voyage. Many of our men were also suffering from fever. The malaria of the dense masses of floating vegetation was most poisonous, and upon looking back to the canoe that followed in our wake I observed all my men sitting crouched together sick and dispirited, looking like departed spirits being ferried across the melancholy Styx.

The woman Bacheeta knew the country, as she had formerly been to Magungo when in the service of Sali, who had been subsequently murdered by Kamrasi. She informed me on the second day that we should terminate our canoe voyage on that day, as we should arrive at the great waterfall of which she had often spoken. As we proceeded the river gradually narrowed to about 180 yards, and when the paddles ceased working we could distinctly hear the roar of water. I had heard this on waking in the morning, but at the time I had imagined it to proceed from distant thunder. By ten o'clock the current had so increased as we proceeded that it was distinctly perceptible, although weak. The roar of the waterfall was extremely loud, and after sharp pulling for a couple of hours, during which time the stream increased, we arrived at a few deserted fishing-huts, at a point where the river made a slight turn. I never saw such an extraordinary show of crocodiles as were exposed on every sandbank on the sides of the river. They lay like logs of timber close together, and upon one bank we counted twenty-seven of large size. Every basking place was crowded in a similar manner. From the time we had fairly entered the river it had been confined by heights somewhat precipitous on either side, rising to about 180 feet. At this point the cliffs were still higher and exceedingly abrupt. From the roar of the water I was sure that the fall would be in sight if we turned the corner at the bend of the river; accordingly I ordered the boatmen to row as far as they could. To this they at first objected, as they wished to stop at the deserted fishing village, which they explained was to be the limit of the journey, further progress being impossible.

However, I explained that I merely wished to see the falls, and they rowed immediately up the stream, which was now strong against us. Upon rounding the corner a magnificent sight burst suddenly upon us. On either side the river were beautifully wooded cliffs rising abruptly to a height of about 300 feet; rocks were jutting out from the intensely green foliage; and rushing through a gap that cleft the rock exactly before us, the river, contracted from a grand stream, was pent up in a narrow gorge of scarcely fifty yards in width. Roaring furiously through the rock-bound pass, it plunged in one leap of about 120 feet perpendicular into a dark abyss below.

The fall of water was snow-white, which had a superb effect as it contrasted with the dark cliffs that walled the river, while the graceful palms of the tropics and wild plantains perfected the beauty of the view. This was the greatest waterfall of the Nile, and in honor of the distinguished President of the Royal Geographical Society I named it the Murchison Falls, as the most important object throughout the entire course of the river.

At this point we had ordered our oxen to he sent, as we could go no farther in the canoes. We found the oxen ready for us; but if we looked wretched, the animals were a match. They had been bitten by the flies, thousands of which were at this spot. Their coats were staring, ears drooping, noses running, and heads hanging down--all the symptoms of fly-bite, together with extreme looseness of the bowels. I saw that it was all up with our animals. Weak as I was myself, I was obliged to walk, as my ox could not carry me up the steep inclination. I toiled languidly to the summit of the cliff, and we were soon above the falls, and arrived at a small village a little before evening.

On the following morning we started, the route as before being parallel to the river, and so close that the roar of the rapids was extremely loud. The river flowed in a deep ravine upon our left. We continued for a day's march along the Somerset, crossing many ravines and torrents, until we turned suddenly down to the left, and arriving at the bank we were to be transported to an island called Patooan, that was the residence of a chief. It was about an hour after sunset, and, being dark, my riding ox, which was being driven as too weak to carry me, fell into an elephant pitfall. After much hallooing, a canoe was brought from the island, which was not more than fifty yards from the mainland, and we were ferried across. We were both very ill with a sudden attack of fever; and my wife, not being able to stand, was, on arrival at the island, carried on a litter I knew not whither, escorted by some of my men, while I lay down on the wet ground quite exhausted with the annihilating disease. At length the rest of my men crossed over, and those who had carried my wife to the village returning with firebrands, I managed to creep after them with the aid of a long stick, upon which I rested with both hands. After a walk through a forest of high trees for about a quarter of a mile, I arrived at a village where I was shown a wretched hut, the stars being visible through the roof. In this my wife lay dreadfully ill upon her angarep, and I fell down upon some straw. About an hour later a violent thunderstorm broke over us, and our hut was perfectly flooded. Being far too ill and helpless to move from our positions, we remained dripping wet and shivering with fever until the morning. Our servants and people had, like all native, made themselves much more comfortable than their employers; nor did they attempt to interfere with our misery in any way until summoned to appear at sunrise.

The island of Patooan was about half a mile long by 150 yards wide, and was one of the numerous masses of rocks that choke the river between Karuma Falls and the great Murchison cataract. My headman now informed me that war was raging between Kamrasi and his rivals, Fowooka and Rionga, and it would be impossible to proceed along the bank of the river to Karuma. My exploration was finished, however, as it was by no means necessary to continue the route from Patooan to Karuma.

CHAPTER XXII.

Prisoners on the island--Left to starve--Months of helplessness-- We rejoin the Turks--The real Kamrasi--In the presence of royalty.

We were prisoners on the island of Patooan as we could not procure porters at any price to remove our effects. We had lost all our riding oxen within a few days. They had succumbed to the flies, and the only animal alive was already half dead; this was the little bull that had always carried the boy Saat. It was the 8th of April, and within a few days the boats upon which we depended for our return to civilization would assuredly quit Gondokoro. I offered the natives all the beads that I had (about 50 lbs.) and the whole of my baggage, if they would carry us to Shooa directly from this spot. We were in perfect despair, as we were both completely worn out with fever and fatigue, and certain death seemed to stare us in the face should we remain in this unhealthy spot. Worse than death was the idea of losing the boats and becoming prisoners for another year in this dreadful land, which must inevitably happen should we not hurry directly to Gondokoro without delay. The natives with their usual cunning at length offered to convey us to Shooa, provided that I paid them the beads in advance. The boats were prepared to ferry us across the river; but I fortunately discovered through the woman Bacheeta their treacherous intention of placing us on the uninhabited wilderness on the north side, and leaving us to die of hunger. They had conspired together to land us, but to return immediately with the boats after having thus got rid of the incubus of their guests.

We were in a great dilemma. Had we been in good health, I would have forsaken everything but the guns and ammunition, and have marched directly to Gondokoro on foot; but this was utterly impossible. Neither my wife nor I could walk a quarter of a mile without fainting. There was no guide, and the country was now overgrown with impenetrable grass and tangled vegetation eight feet high. We were in the midst of the rainy season-- not a day passed without a few hours of deluge. Altogether it was a most heart-breaking position. Added to the distress of mind at being thus thwarted, there was also a great scarcity of provision. Many of my men were weak, the whole party having suffered much from fever; in fact, we were completely helpless.

Our guide, Rabonga, who had accompanied us from M'rooli, had absconded, and we were left to shift for ourselves. I was determined not to remain on the island, as I suspected that the boats might be taken away, and that we should be kept prisoners; I therefore ordered my men to take the canoes, and to ferry us to the main land, from whence we had come. The headman, upon hearing this order, offered to carry us to a village, and then to await orders from Kamrasi as to whether we were to be forwarded to Shooa or not. The district in which the island of Patooan was situated was called Shooa Moru, although having no connection with the Shooa in the Madi country to which we were bound.

We were ferried across to the main shore, and my wife and I, in our respective angareps, were carried by the natives for about three miles. Arriving at a deserted village, half of which was in ashes, having been burned and plundered by the enemy, we were deposited on the ground in front of an old hut in the pouring rain, and were informed that we should remain there that night, but that on the following morning we should proceed to our destination.

Not trusting the natives, I ordered my men to disarm them, and to retain their spears and shields as security for their appearance on the following day. This effected, we were carried into a filthy hut about six inches deep in mud, as the roof was much out of repair, and the heavy rain had flooded it daily for some weeks. I had a canal cut through the muddy floor, and in misery and low spirits we took possession.

On the following morning not a native was present! We had been entirely deserted; although I held the spears and shields, every man had absconded. There were neither inhabitants nor provisions. The whole country was a wilderness of rank grass that hemmed us in on all sides. Not an animal, nor even a bird, was to be seen; it was a miserable, damp, lifeless country. We were on elevated ground, and the valley of the Somerset was about two miles to our north, the river roaring sullenly in its obstructed passage, its course marked by the double belt of huge dark trees that grew upon its banks.

My men naturally felt outraged and proposed that we should return to Patooan, seize the canoes, and take provisions by force, as we had been disgracefully deceived. The natives had merely deposited us here to get us out of the way, and in this spot we might starve. Of course I would not countenance the proposal of seizing provisions, but I directed my men to search among the ruined villages for buried corn, in company with the woman Bacheeta, who, being a native of this country, would be up to the ways of the people, and might assist in the discovery.

After some hours passed in rambling over the black ashes of several villages that had been burned, they discovered a hollow place, by sounding the earth with a stick, and, upon digging, arrived at a granary of the seed known as "tullaboon;" this was a great prize, as, although mouldy and bitter, it would keep us from starving. The women of the party were soon hard at work grinding, as many of the necessary stones had been found among the ruins.

Fortunately there were three varieties of plants growing wild in great profusion, that, when boiled, were a good substitute for spinach; thus we were rich in vegetables, although without a morsel of fat or animal food. Our dinner consisted daily of a mess of black porridge of bitter mouldy flour that no English pig would condescend to notice, and a large dish of spinach. "Better a dinner of herbs where love is," etc. often occurred to me; but I am not sure that I was quite of that opinion after a fortnight's grazing upon spinach.

Tea and coffee were things of the past, the very idea of which made our months water; but I found a species of wild thyme growing in the jungles, and this when boiled formed a tolerable substitute for tea. Sometimes our men procured a little wild honey, which added to the thyme tea we considered a great luxury.

This wretched fare, in our exhausted state from fever and general effects of climate, so completely disabled us that for nearly two months my wife lay helpless on one angarep, and I upon the other. Neither of us could walk. The hut was like all in Kamrasi's country, with a perfect forest of thick poles to support the roof (I counted thirty-two); thus, although it was tolerably large, there was but little accommodation. These poles we now found very convenient, as we were so weak that we could not rise from bed without lifting ourselves up by one of the supports.

We were very nearly dead, and our amusement was a childish conversation about the good things in England, and my idea of perfect happiness was an English beefsteak and a bottle of pale ale; for such a luxury I would most willingly have sold my birthright at that hungry moment. We were perfect skeletons, and it was annoying to see how we suffered upon the bad fare, while our men apparently throve. There were plenty of wild red peppers, and the men seemed to enjoy a mixture of porridge and legumes a la sauce piquante. They were astonished at my falling away on this food, but they yielded to my argument when I suggested that a "lion would starve where a donkey grew fat." I must confess that this state of existence did not improve my temper, which, I fear, became nearly as bitter as the porridge. My people had a windfall of luck, as Saat's ox, that had lingered for a long time, lay down to die, and stretching himself out, commenced kicking his last kick. The men immediately assisted him by cutting his throat, and this supply of beef was a luxury which, even in my hungry state, was not the English beefsteak for which I sighed, and I declined the diseased bull.

The men made several long excursions through the country to purchase provisions, but in two months they procured only two kids; the entire country was deserted, owing to the war between Kamrasi and Fowooka. Every day the boy Saat and the woman Bacheeta sallied out and conversed with the inhabitants of the different islands on the river. Sometimes, but very rarely, they returned with a fowl; such an event caused great rejoicing.

We gave up all hope of Gondokoro, and were resigned to our fate. This, we felt sure, was to be buried in Chopi, the name of our village. I wrote instructions in my journal, in case of death, and told my headman to be sure to deliver my maps, observations, and papers to the English Consul at Khartoum. This was my only care, as I feared that all my labor might be lost should I die. I had no fear for my wife, as she was quite as bad as I, and if one should die the other would certainly follow; in fact, this had been agreed upon, lest she should fall into the hands of Kamrasi at my death. We had struggled to win, and I thanked God that we had won. If death were to be the price, at all events we were at the goal, and we both looked upon death rather as a pleasure, as affording REST. There would be no more suffering, no fever, no long journey before us, that in our weak state was an infliction. The only wish was to lay down the burden. Curious is the warfare between the animal instincts and the mind! Death would have been a release that I would have courted; but I should have liked that one "English beefsteak and pale ale" before I died!

During our misery of constant fever and starvation at Shooa Moru, insult had been added to injury. There was no doubt that we had been thus deserted by Kamrasi's orders, as every seven or eight days one of his chiefs arrived and told me that the king was with his army only four days' march from me, and that he was preparing to attack Fowooka, but that he wished me to join him, as with my fourteen guns, we should win a great victory. This treacherous conduct, after his promise to forward me without delay to Shooa, enraged me exceedingly. We had lost the boats at Gondokoro, and we were now nailed to the country for another year, should we live, which was not likely. Not only had the brutal king thus deceived us, but he was deliberately starving us into conditions, his aim being that my men should assist him against his enemy. At one time the old enemy tempted me sorely to join Fowooka against Kamrasi; but, discarding the idea, generated in a moment of passion, I determined to resist his proposals to the last. It was perfectly true that the king was within thirty miles of us, that he was aware of our misery, and made use of our extremity to force us to become his allies.

After more than two months passed in this distress it became evident that something must be done. I sent my headman, or vakeel, and one man, with a native as a guide (that Saat and Bacheeta had procured from an island), with instructions to go direct to Kamrasi, to abuse him thoroughly in my name for having thus treated us, and tell him that I was much insulted at his treating with me through a third party in proposing an alliance. My vakeel was to explain that I was a much more powerful chief than Kamrasi, and that if he required my alliance, he must treat with me in person, and immediately send fifty men to transport my wife, myself, and effects to his camp, where we might, in a personal interview, come to terms.

I told my vakeel to return to me with the fifty men, and to be sure to bring from Kamrasi some token by which I should know that he had actually seen him. The vakeel and Yaseen started.

After some days the absconded guide, Rabonga, appeared with a number of men, but without either my vakeel or Yaseen. He carried with him a small gourd bottle, carefully stopped; this he broke, and extracted from the inside two pieces of printed paper that Kamrasi had sent to me in reply.

On examining the papers, I found them to be portions of the English Church Service translated into (I think) the "Kisuabili" language, by Dr Krapf! There were many notes in pencil on the margin, written in English, as translations of words in the text. It quickly occurred to me that Speke must have given this book to Kamrasi on his arrival from Zanzibar, and that he now extracted the leaves and sent them to me as a token I had demanded to show that my message had been delivered to him.

Rabonga made a lame excuse for his previous desertion. He delivered a thin ox that Kamrasi had sent me, and he declared that his orders were that he should take my whole party immediately to Kamrasi, as he was anxious that we should attack Fowooka without loss of time. We were positively to start on the following morning! My bait had taken, and we should escape from this frightful spot, Shooa Moru.

After winding through dense jungles of bamboos and interminable groves of destroyed plantains, we perceived the tops of a number of grass hats appearing among the trees. My men now begged to be allowed to fire a salute, as it was reported that the ten men of Ibrahim's party who had been left as hostages were quartered at this village with Kamrasi. Hardly had the firing commenced when it was immediately replied to by the Turks from their camp, who, upon our approach, came out to meet us with great manifestations of delight and wonder at our having accomplished our long and difficult voyage.

My vakeel and Yaseen were the first to meet us, with an apology that severe fever had compelled them to remain in camp instead of returning to Shooa Moru according to my orders; but they had delivered my message to Kamrasi, who had, as I had supposed, sent two leaves out of a book Speke had given him, as a reply. An immense amount of news had to be exchanged between my men and those of Ibrahim. They had quite given us up for lost, until they heard that we were at Shooa Moru. A report had reached them that my wife was dead, and that I had died a few days later. A great amount of kissing and embracing took place, Arab fashion, between the two parties; and they all came to kiss my hand and that of my wife, with the exclamation, that "By Allah, no woman in the world had a heart so tough as to dare to face what she had gone through." "El hamd el Illah! El hamd el Illah bel salaam!" ("Thank God--be grateful to God") was exclaimed on all sides by the swarthy throng of brigands who pressed round us, really glad to welcome us back again; and I could not help thinking of the difference in their manner now and fourteen months before, when they had attempted to drive us back from Gondokoro.

Hardly were we seated in our hut when my vakeel announced that Kamrasi had arrived to pay me a visit. In a few minutes he was ushered into the hut. Far from being abashed, he entered with a loud laugh, totally different from his former dignified manner. "Well, here you are at last!" he exclaimed. Apparently highly amused with our wretched appearance, he continued, "So you have been to the M'wootan N'zige! Well, you don't look much the better for it; why, I should not have known you! ha, ha, ha!" I was not in a humor to enjoy his attempts at facetiousness; I therefore told him that he had behaved disgracefully and meanly, and that I should publish his character among the adjoining tribes as below that of the most petty chief that I had ever seen.

"Never mind," he replied, "it's all over now." You really are thin, both of you. It was your own fault; why did you not agree to fight Fowooka? You should have been supplied with fat cows and milk and butter, had you behaved well.

I will have my men ready to attack Fowooka to-morrow. The Turks have ten men, you have thirteen; thirteen and ten make twenty-three. You shall be carried if you can't walk, and we will give Fowooka no chance. He must be killed--only kill him, and MY BROTHER will give you half of his kingdom."

He continued, "You shall have supplies to-morrow; I will go to my BROTHER, who is the great M'Kamma Kamrasi, and he will send you all you require. I am a little man; he is a big one. I have nothing; he has everything, and he longs to see you. You must go to him directly; he lives close by."

I hardly knew whether he was drunk or sober. "My bother the great M'Kamma Kamrasi!" I felt bewildered with astonishment. Then, "If you are not Kamrasi, pray who are you?" I asked. "Who am I?" he replied. "Ha, ha, ha! that's very good; who am I?-- I am M'Gambi, the brother of Kamrasi; I am the younger brother, but HE IS THE KING."

The deceit of this country was incredible. I had positively never seen the real Kamrasi up to this moment, and this man M'Gambi now confessed to having impersonated the king, his brother, as Kamrasi was afraid that I might be in league with Debono's people to murder him, and therefore he had ordered his brother M'Gambi to act the king.

I told M'Gambi that I did not wish to see his brother, the king, as I should perhaps be again deceived and be introduced to some impostor like himself; and that as I did not choose to be made a fool of, I should decline the introduction. This distressed him exceedingly. He said that the king was really so great a man that he, his own brother, dared not sit on a stool in his presence, and that he had only kept in retirement as a matter of precaution, as Debono's people had allied themselves with his enemy Rionga in the preceding year, and he dreaded treachery. I laughed contemptuously at M'Gambi, telling him that if a woman like my wife dared to trust herself far from her own country among such savages as Kamrasi's people, their king must be weaker than a woman if he dared not show himself in his own territory. I concluded by saying that I should not go to see Kamrasi, but that he should come to visit me.

On the following morning, after my arrival at Kisoona, M'Gambi appeared, beseeching me to go and visit the king. I replied that "I was hungry and weak from want of food, and that I wanted to see meat, and not the man who had starved me." In the afternoon a beautiful cow appeared with her young calf, also a fat sheep and two pots of plantain cider, as a present from Kamrasi. That evening we revelled in milk, a luxury that we had not tasted for some months. The cow gave such a quantity that we looked forward to the establishment of a dairy, and already contemplated cheese-making. I sent the king a present of a pound of powder in canister, a box of caps, and a variety of trifles, explaining that I was quite out of stores and presents, as I had been kept so long in his country that I was reduced to beggary, as I had expected to return to my own country long before this.

In the evening M'Gambi appeared with a message from the king, saying that I was his greatest friend, and that he would not think of taking anything from me as he was sure that I must be hard up; that he desired nothing, but would be much obliged if I would give him the "little double rifle that I always carried, and my watch and compass!" He wanted "NOTHING," only my Fletcher rifle, that I would as soon have parted with as the bone of my arm; and these three articles were the same for which I had been so pertinaciously bored before my departure from M'rooli. It was of no use to be wroth, I therefore quietly replied that I should not give them, as Kamrasi had failed in his promise to forward me to Shooa; but that I required no presents from him, as he always expected a thousandfold in return. M'Gambi said that all would be right if I would only agree to pay the king a visit. I objected to this, as I told him the king, his brother, did not want to see me, but only to observe what I had, in order to beg for all that he saw. He appeared much hurt, and assured me that he would be himself responsible that nothing of the kind should happen, and that he merely begged as a favor that I would visit the king on the following morning, and that people should be ready to carry me if I were unable to walk. Accordingly I arranged to be carried to Kamrasi's camp at about 8 A.M.

At the hour appointed M'Gambi appeared, with a great crowd of natives. My clothes were in rags, and as personal appearance has a certain effect, even in Central Africa, I determined to present myself to the king in as favorable a light as possible. I happened to possess a full-dress Highland suit that I had worn when I lived in Perthshire many years before. This I had treasured as serviceable upon an occasion like the present: accordingly I was quickly attired in kilt, sporran, and Glengarry bonnet, and to the utter amazement of the crowd, the ragged-looking object that had arrived in Kisoona now issued from the obscure hut with plaid and kilt of Athole tartan. A general shout of exclamation arose from the assembled crowd, and taking my seat upon an angarep, I was immediately shouldered by a number of men, and, attended by ten of my people as escort, I was carried toward the camp of the great Kamrasi.

In about half an hour we arrived. The camp, composed of grass huts, extended over a large extent of ground, and the approach was perfectly black with the throng that crowded to meet me. Women, children, dogs, and men all thronged at the entrance of the street that led to Kamrasi's residence. Pushing our way through this inquisitive multitude, we continued through the camp until at length we reached the dwelling of the king. Halting for the moment, a message was immediately received that we should proceed; we accordingly entered through a narrow passage between high reed fences, and I found myself in the presence of the actual king of Unyoro, Kamrasi. He was sitting in a kind of porch in front of a hut, and upon seeing me he hardly condescended to look at me for more than a moment; he then turned to his attendants and made some remark that appeared to amuse them, as they all grinned as little men are wont to do when a great man makes a bad joke.

I had ordered one of my men to carry my stool; I was determined not to sit upon the earth, as the king would glory in my humiliation. M'Gambi, his brother, who had formerly played the part of king, now sat upon the ground a few feet from Kamrasi, who was seated upon the same stool of copper that M'Gambi had used when I first saw him at M'rooli. Several of his chiefs also sat upon the straw with which the porch was littered. I made a "salaam" and took my seat upon my stool.

Not a word passed between us for about five minutes, during which time the king eyed me most attentively, and made various remarks to the chiefs who were present. At length he asked me why I had not been to see him before. I replied, because I had been starved in his country, and I was too weak to walk. He said I should soon be strong, as he would now give me a good supply of food; but that he could not send provisions to Shooa Moru, as Fowooka held that country. Without replying to this wretched excuse for his neglect, I merely told him that I was happy to have seen him before my departure, as I was not aware until recently that I had been duped by

M'Gambi. He answered me very coolly, saying that although I had not seen him, he had nevertheless seen me, as he was among the crowd of native escort on the day that we left M'rooli. Thus he had watched our start at the very place where his brother M'Gambi had impersonated the king.

Kamrasi was a remarkably fine man, tall and well proportioned, with a handsome face of a dark brown color, but a peculiarly sinister expression. He was beautifully clean, and instead of wearing the bark cloth common among the people, he was dressed in a fine mantle of black and white goatskins, as soft as chamois leather. His people sat on the ground at some distance from his throne; when they approached to address him on any subject they crawled upon their hands and knees to his feet, and touched the ground with their foreheads.

True to his natural instincts, the king commenced begging, and being much struck with the Highland costume, he demanded it as a proof of friendship, saying that if I refused I could not be his friend. The watch, compass, and double Fletcher rifle were asked for in their turn, all of which I refused to give him. He appeared much annoyed, therefore I presented him with a pound canister of powder, a box of caps, and a few bullets. He asked, "What's the use of the ammunition if you won't give me your rifle?" I explained that I had already given him a gun, and that he had a rifle of Speke's. Disgusted with his importunity I rose to depart, telling him that I should not return to visit him, as I did not believe he was the real Kamrasi I had heard that Kamrasi was a great king, but he was a mere beggar, and was doubtless an impostor, like M'Gambi. At this he seemed highly amused, and begged me not to leave so suddenly, as he could not permit me to depart empty-handed. He then gave certain orders to his people, and after a little delay two loads of flour arrived, together with a goat and two jars of sour plantain cider. These presents he ordered to be forwarded to Kisoona. I rose to take leave; but the crowd, eager to see what was going forward, pressed closely upon the entrance of the approach, seeing which, the king gave certain orders, and immediately four or five men with long heavy bludgeons rushed at the mob and belabored them right and left, putting the mass to flight pell-mell through the narrow lanes of the camp.

I was then carried back to my camp at Kisoona, where I was received by a great crowd of people.

CHAPTER XXIII.

The hour of deliverance--Triumphal entry into Gondokoro--Home-bound--The plague breaks out--Our welcome at Khartoum to civilization.

The hour of deliverance from our long sojourn in Central Africa was at hand. It was the month of February, and the boats would be at Gondokoro. The Turks had packed their ivory; the large tusks were fastened to poles to be carried by two men, and the camp was a perfect mass of this valuable material. I counted 609 loads of upward of 50 lbs. each; thirty-one loads were lying at an out-station; therefore the total results of the ivory campaign during the last twelve months were about 32,000 lbs., equal to about 9,630 pounds sterling when delivered in Egypt. This was a perfect fortune for Koorshid.

We were ready to start. My baggage was so unimportant that I was prepared to forsake everything, and to march straight for Gondokoro independently with my own men; but this the Turks assured me was impracticable, as the country was so hostile in advance that we must of necessity have some fighting on the road; the Bari tribe would dispute our right to pass through their territory.

The day arrived for our departure; the oxen were saddled, and we were ready to start. Crowds of people cane to say "good-by;" but, dispensing with the hand-kissing of the Turks who were to remain in camp, we prepared for our journey toward HOME. Far away though it was, every step would bring us nearer. Nevertheless there were ties even in this wild spot, where all was savage and unfeeling--ties that were painful to sever, and that caused a sincere regret to both of us when we saw our little flock of unfortunate slave children crying at the idea of separation. In this moral desert, where all humanized feelings were withered and parched like the sands of the Soudan, the guilelessness of the children had been welcomed like springs of water, as the only refreshing feature in a land of sin and darkness.

"Where are you going?" cried poor little Abbai in the broken Arabic that we had taught him. "Take me with you, Sitty!" (lady), and he followed us down the path, as we regretfully left our proteges, with his fists tucked into his eyes, weeping from his heart, although for his own mother he had not shed a tear. We could not take him with us; he belonged to Ibrahim, and had I purchased the child to rescue him from his hard lot and to rear him as a civilized being, I might have been charged with slave-dealing. With heavy hearts we saw hint taken up in the arms of a woman and carried back to camp, to prevent him from following our party, that had now started.

I will not detain the reader with the details of our journey home. After much toil and some fighting with hostile natives, we bivouacked one sunset three miles from Gondokoro. That night we were full of speculations. Would a boat be waiting for us with supplies and letters? The morning anxiously looked forward to at length arrived. We started. The English flag had been mounted on a fine straight bamboo with a new lance-head specially arranged for the arrival at Gondokoro. My men felt proud, as they would march in as conquerors. According to White Nile ideas, such a journey could not have been accomplished with so small a party. Long before Ibrahim's men were ready to start, our oxen were saddled and we were off, longing to hasten into Gondokoro and to find a comfortable vessel with a few luxuries and the post from England. Never had the oxen travelled so fast as on that morning; the flag led the way, and the men, in excellent spirits, followed at double-quick pace.

"I see the masts of the vessels!" exclaimed the boy Saat. "El hambd el Illah!" (Thank God!) shouted the men. "Hurrah!" said I; "Three cheers for Old England and the Sources of the Nile! Hurrah!" and my men joined me in the wild, and to their ears savage, English yell. "Now for a salute! Fire away all your powder, if you like, my lads, and let the people know that we're alive!"

This was all that was required to complete the happiness of my people, and, loading and firing as fast as possible, we approached near to Gondokoro. Presently we saw the Turkish flag emerge from Gondokoro at about a quarter of a mile distant, followed by a number of the traders' people, who waited to receive us. On our arrival they immediately approached and fired salutes with ball cartridge, as usual advancing close to us and discharging their guns into the ground at our feet. One of my servants, Mahomet, was riding an ox, and an old friend of his in the crowd happening to recognize him immediately advanced and saluted him by firing his gun into the earth directly beneath the belly of the ox he was riding.

The effect produced made the crowd and ourselves explode with laughter. The nervous ox, terrified at the sudden discharge between his legs, gave a tremendous kick, and continued madly kicking and plunging, until Mahomet was pitched over his head and lay sprawling on the ground. This scene terminated the expedition.

Dismounting from our tired oxen, our first inquiry was concerning boats and letters. What was the reply? Neither boats, letters, supplies, nor any intelligence of friends or the civilized world! We had long since been given up as dead by the inhabitants of Khartoum, and by all those who understood the difficulties and dangers of the country. We were told that some people had suggested that we might possibly have gone to Zanzibar, but the general opinion was that we had all been killed.

At this cold and barren reply I felt almost choked. We had looked forward to arriving at Gondokoro as to a home; we had expected that a boat would have been sent on the chance of finding us, as I had left money in the hands of an agent in Khartoum ; but there was literally nothing to receive us, and we were helpless to return. We had worked for years in misery, such as I have but faintly described, to overcome the difficulties of this hitherto unconquerable exploration. We had succeeded--and what was the result? Not even a letter from home to welcome us if alive!

As I sat beneath a tree and looked down upon the glorious Nile that flowed a few yards beneath my feet, I pondered upon the value of my toil. I had traced the river to its great Albert source, and as the mighty stream glided before me, the mystery that had ever shrouded its origin was dissolved. I no longer looked upon its waters with a feeling approaching to awe, for I knew its home, and had visited its cradle. Had I overrated the importance of the discovery? and had I wasted some of the best years of my life to obtain a shadow? I recalled to recollection the practical question of Commoro, the chief of Latooka, "Suppose you get to the great lake, what will you do with it? What will be the good of it? If you find that the large river does flow from it, what then?"

At length the happy day came when we were to quit this miserable place of Gondokoro. The boat was ready to start, we were all on board, and Ibrahim and his people came to say good-by. Crowds lined the cliff and the high ground by the old ruins of the mission-station to see us depart. We pushed off from shore into the powerful current; the English flag, that had accompanied us all through our wanderings, now fluttered proudly from the masthead unsullied by defeat, and amidst the rattle of musketry we glided rapidly down the river and soon lost sight of Gondokoro.

What were our feelings at that moment? Overflowing with gratitude to a Divine Providence that had supported us in sickness and guided us through all dangers. There had been moments of hopelessness and despair; days of misery, when the future had appeared dark and fatal; but we had been strengthened in our weakness, and led, when apparently lost, by an unseen hand. I felt no triumph, but with a feeling of calm contentment and satisfaction we floated down the Nile. My great joy was in the meeting that I contemplated with Speke in England, as I had so thoroughly completed the task we had agreed upon.

We had heard at Gondokoro of a remarkable obstruction in the White Nile a short distance below the junction of the Bahr el Gazal. We found this to be a dam formed by floating masses of vegetation that effectually blocked the passage.

The river had suddenly disappeared; there was apparently an end to the White Nile. The dam was about three-quarters of a mile wide, was perfectly firm, and was already overgrown with high reeds and grass, thus forming a continuation of the surrounding country. Many of the traders' people had died of the plague at this spot during the delay of some weeks in cutting the canal; the graves of these dead were upon the dam. The bottom of the canal that had been cut through the dam was perfectly firm, composed of sand, mud, and interwoven decaying vegetation. The river arrived with great force at the abrupt edge of the obstruction, bringing with it all kinds of trash and large floating islands. None of these objects hitched against the edge, but the instant they struck they dived under and disappeared. It was in this manner that a vessel had recently been lost. Having missed the narrow entrance to the canal, she had struck the dam stem on; the force of the current immediately turned her broadside against the obstruction, the floating islands and masses of vegetation brought down by the river were heaped against her and, heeling over on her side, she was sucked bodily under and carried beneath the dam. Her crew had time to save themselves by leaping upon the firm barrier that had wrecked their ship. The boatmen told me that dead hippopotami had been found on the other side, that had been carried under the dam and drowned.

Two days' hard work from morning till night brought us through the canal, and we once more found ourselves on the open Nile on the other side of the dam. The river was in that spot perfectly clean; not a vestige of floating vegetation could be seen upon its waters. In its subterranean passage it had passed through a natural sieve, leaving all foreign matter behind to add to the bulk of the already stupendous work.

All before us was clear and plain sailing. For some days two or three of our men had been complaining of severe headache, giddiness, and violent pains in the spine and between the shoulders. I had been anxious when at Gondokoro concerning the vessel, as many persons while on board had died of the plague, during the voyage from Khartoum. The men assured me that the most fatal symptom was violent bleeding from the nose; in such cases no one had been known to recover. One of the boatmen, who had been ailing for some days, suddenly went to the side of the vessel and hung his head over the river; his nose was bleeding!

Another of my men, Yaseen, was ill; his uncle, my vakeel, came to me with a report that "his nose was bleeding violently!" Several other men fell ill; they lay helplessly about the deck in low muttering delirium, their eyes as yellow as orange-peel. In two or three days the vessel was so horribly offensive as to be unbearable. THE PLAGUE HAD BROKEN OUT! We floated past the river Sobat junction; the wind was fair from the south, thus fortunately we in the stern were to windward of the crew. Yaseen died; he was one who had bled at the nose. We stopped to bury him. The funeral hastily arranged, we again set sail. Mahommed died; he had bled at the nose. Another burial. Once more we set sail and hurried down the Nile. Several men were ill, but the dreaded symptom had not appeared. I had given each man a strong dose of calomel at the commencement of the disease; I could do nothing more, as my medicines were exhausted. All night we could hear the sick muttering and raving in delirium, but from years of association with disagreeables we had no fear of the infection.

One morning the boy Saat came to me with his head bound up, and complained of severe pain in the back and limbs, with all the usual symptoms of plague. In the afternoon I saw him leaning over the ship's side; his nose was bleeding violently! At night he was delirious. On the following morning he was raving, and on the vessel stopping to collect firewood he threw himself into the river to cool the burning fever that consumed him. His eyes were suffused with blood, which, blended with a yellow as deep as the yolk of egg, gave a terrible appearance to his face, that was already so drawn and changed as to be hardly recognized. Poor Saat! the faithful boy that we had adopted, and who had formed so bright an exception to the dark character of his race, was now a victim to this horrible disease. He was a fine strong lad of nearly fifteen, and he now lay helplessly on his mat, and cast wistful glances at the face of his mistress as she gave him a cup of cold water mixed with a few lumps of sugar that we had obtained from the traders at Gondokoro.

Saat grew worse and worse. Nothing would relieve the unfortunate boy from the burning torture of that frightful disease. He never slept; but night and day he muttered in delirium, breaking the monotony of his malady by occasionally howling like a wild animal. Richarn won my heart by his careful nursing of the boy, who had been his companion through years of hardship. We arrived at the village of Wat Shely, only three days from Khartoum. Saat was dying. The night passed, and I expected that all would be over before sunrise; but as morning dawned a change had taken place; the burning fever had left him, and, although raised blotches had broken out upon his chest and various parts of his body, he appeared much better. We now gave him stimulants; a teaspoonful of araki that we had bought at Fashooder was administered every ten minutes on a lump of sugar. This he crunched in his mouth, while he gazed at my wife with an expression of affection; but he could not speak. I had him well washed and dressed in clean clothes, that had been kept most carefully during the voyage, to be worn on our entree to Khartoum. He was laid down to sleep upon a clean mat, and my wife gave him a lump of sugar to moisten his mouth and relieve his thickly-furred tongue. His pulse was very weak, and his skin cold. "Poor Saat," said my wife, "his life hangs upon a thread. We must nurse him most carefully; should he have a relapse, nothing will save him."

An hour passed, and he slept. Karka, the fat, good-natured slave woman, quietly went to his side; gently taking him by the ankles and knees, she stretched his legs into a straight position, and laid his arms parallel with his sides. She then covered his face with a cloth, one of the few rags that we still possessed. "Does he sleep still?" we asked. The tears ran down the cheeks of the savage but good-hearted Karka as she sobbed, "He is dead!"

We stopped the boat. It was a sandy shore; the banks were high, and a clump of mimosas grew above high-water mark. It was there that we dug his grave. My men worked silently and sadly, for all loved Saat. He had been so good and true, that even their hard hearts had learned to respect his honesty. We laid him in his grave on the desert shore, beneath the grove of trees.

Again the sail was set, and, filled by the breeze, it carried us away from the dreary spot where we had sorrowfully left all that was good and faithful. It was a happy end-- most merciful, as he had been taken from a land of iniquity in all the purity of a child converted from Paganism to Christianity. He had lived and died in our service a good Christian. Our voyage was nearly over, and we looked forward to home and friends; but we had still fatigues before us: poor Saat had reached his home and rest.

On the following morning, May 6, 1865, we were welcomed by the entire European population of Khartoum, to whom are due my warmest thanks for many kind attentions. We were kindly offered a house by Monsieur Lombrosio, the manager of the Khartoum branch of the "Oriental and Egyptian Trading Company."

I now heard the distressing news of the death of my poor friend Speke. I could not realize the truth of this melancholy report until I read the details of his fatal accident in the appendix of a French translation of his work. It was but a sad consolation that I could confirm his discoveries, and bear witness to the tenacity and perseverance with which he had led his party through the untrodden path of Africa to the first Nile source.

While at Khartoum I happened to find Mahommed Iler! the vakeel of Chenooda's party, who had instigated my men to mutiny at Latooka, and had taken my deserters into his employ. I had promised to make an example of this fellow; I therefore had him arrested and brought before the divan. With extreme effrontery, he denied having had anything to do with the affair. Having a crowd of witnesses in my own men, and others that I had found in

Khartoum who had belonged to Koorshid's party at that time, his barefaced lie was exposed, and he was convicted. I determined that he should be punished, as an example that would insure respect to any future English traveller in those regions. My men, and all those with whom I had been connected, had been accustomed to rely most implicitly upon all that I had promised, and the punishment of this man had been an expressed determination.

I went to the divan and demanded that he should be flogged. Omer Bey was then Governor of the Soudan, in the place of Moosa Pacha deceased. He sat upon the divan, in the large hall of justice by the river. Motioning me to take a seat by his side, and handing me his pipe, he called the officer in waiting, and gave the necessary orders. In a few minutes the prisoner was led into the hall, attended by eight soldiers. One man carried a strong pole about seven feet long, in the centre of which was a double chain, riveted through in a loop. The prisoner was immediately thrown down with his face to the ground, while two men stretched out his arms and sat upon them. His feet were then placed within the loop of the chain, and the pole being twisted round until firmly secured, it was raised from the ground sufficiently to expose the soles of the feet. Two men with powerful hippopotamus whips stood one on either side. The prisoner thus secured, the order was given. The whips were most scientifically applied, and after the first five dozen the slave-hunting scoundrel howled most lustily for mercy. How often had he flogged unfortunate slave women to excess, and what murders had that wretch committed, who now howled for mercy! I begged Omer Bey to stop the punishment at 150 lashes, and to explain to him publicly in the divan that he was thus punished for attempting to thwart the expedition of an English traveller, by instigating my escort to mutiny.

We stayed at Khartoum two months, waiting for the Nile to rise sufficiently to allow the passage of the cataracts. We started June 30th, and reached Berber, from which point, four years before, I had set out on my Atbara expedition.

I determined upon the Red Sea route to Egypt, instead of passing the horrible Korosko desert during the hot month of August. After some delay I procured camels, and started east for Souakim, where I hoped to procure a steamer to Suez.

There was no steamer upon our arrival. After waiting in intense heat for about a fortnight, the Egyptian thirty-two-gun steam frigate Ibrahimeya arrived with a regiment of Egyptian troops, under Giaffer Pacha, to quell the mutiny of the black troops at Kassala, twenty days' march in the interior. Giaffer Pacha most kindly placed the frigate at our disposal to convey us to Suez.

Orders for sailing had been received; but suddenly a steamer was signalled as arriving. This was a transport, with troops. As she was to return immediately to Suez, I preferred the dirty transport rather than incur a further delay. We started from Souakim, and after five days' voyage we arrived at Suez. Landing from the steamer, I once more found myself in an English hotel.

The hotel was thronged with passengers to India, with rosy, blooming English ladies and crowds of my own countrymen. I felt inclined to talk to everybody. Never was I so in love with my own countrymen and women; but they (I mean the ladies) all had large balls of hair at the backs of their heads! What an extraordinary change! I called Richarn, my pet savage from the heart of Africa, to admire them. "Now, Richarn, look at them!" I said. "What do you think of the English ladies? eh, Richarn? Are they not lovely?"

"Wah Illahi!" exclaimed the astonished Richarn, "they are beautiful! What hair! They are not like the negro savages, who work other people's hair into their own heads; theirs is all real--all their own--how beautiful!"

"Yes, Richarn," I replied, "ALL THEIR OWN!" This was my first introduction to the "chignon."

We arrived at Cairo, and I established Richarn and his wife in a comfortable situation as private servants to Mr. Zech, the master of Sheppard's Hotel. The character I gave him was one that I trust has done him service. He had shown an extraordinary amount of moral courage in totally reforming from his original habit of drinking. I left my old servant with a heart too full to say good-by, a warm squeeze of his rough but honest black hand, and the whistle of the train sounded-- we were off!

I had left Richarn, and none remained of my people. The past appeared like a dream; the rushing sound of the train renewed ideas of civilization. Had I really come from the Nile Sources? It was no dream. A witness sat before me--a face still young, but bronzed like an Arab by years of exposure to a burning sun, haggard and worn with toil and sickness, and shaded with cares happily now past, the devoted companion of my pilgrimage, to whom I owed success and life-- my wife.

I had received letters from England, that had been waiting at the British Consulate. The first I opened informed me that the Royal Geographical Society had awarded me the Victoria Gold Medal, at a time when they were unaware whether I was alive or dead, and when the success of my expedition was unknown. This appreciation of my exertions was the warmest welcome that I could have received on my first entrance into civilization after so many years of savagedom. It rendered the completion of the Nile Sources doubly grateful, as I had fulfilled the expectations that the Geographical Society had so generously expressed by the presentation of their medal BEFORE my task was done.

###

Printed in Great Britain
by Amazon